MICHIGAN

Rocks & Minerals

A Field Guide to the Great Lake State

Dan R. Lynch & Bob Lynch

Adventure Publications

Cambridge, Minnesota

Dedication

To Nancy Lynch, wife of Bob and mother of Dan, for her continued support of our book projects.

And to Julie Kirsch, Dan's wife, for her infinite patience.

Acknowledgments

Thanks to the following for providing specimens and/or information: George Robinson, Ph.D., Michael P. Basal, John Perona, Richard Whiteman, Ken Flood, Steve Whelan, Karen Brzys, Alex Fagotti, Kennecott Eagle Minerals Company, Daniel R. Fountain and Shawn M. Carlson at the Department of Geography, Northern Michigan University.

Photography by Dan R. Lynch

Cover and book design by Jonathan Norberg

Edited by Brett Ortler

20 19 18 17 16 15 14

Michigan Rocks & Minerals: A Field Guide to the Great Lake State
Copyright © 2010 by Dan R. Lynch and Bob Lynch
Published by Adventure Publications
An imprint of AdventureKEEN
310 Garfield Street South
Cambridge, Minnesota 55008
(800) 678-7006
www.adventurepublications.net
All rights reserved
Printed in China
ISBN 978-1-59193-239-0 (pbk.); ISBN 978-1-59193-666-4 (ebook)

Table of Contents

Glossary Note

We know that books about geology, rocks and minerals can be quite technical. To make this book intuitive for amateurs and useful for professionals, we've included technical terms in the text, but we "translate" the phrases immediately after using them by providing a brief definition. And, of course, all of the geology-related terms we've used are also defined in the glossary found at the back of this book.

For those who are entirely new to rock and mineral collecting, there are a few very important terms you should understand not only before you begin collecting, but even before you read this book: A **crystal** is a solid object with a repeating structure created by a chemical compound. For example, silicon dioxide is a chemical compound consisting of one silicon atom and two oxygen atoms, and when they come together and harden, they form quartz, a white mineral that has six-sided crystals. A "repeating structure" means that when a crystal grows, it builds upon itself. If you have two crystals of quartz, one an inch long while the other is ten feet long, they would have an identical six-sided shape. If a mineral is not found in a well-crystallized form but rather as a solid, rough chunk, it is said to be **massive**. If something forms **massively**, it will always be found as irregular pieces or chunks, rather than as crystals. **Cleavage** is the property of a mineral to break along its molecular crystal planes when carefully struck. A good way to think of cleavage is to imagine a brick wall as being a crystal—if you smash the wall with a wrecking ball, you'll get many broken pieces. But if you carefully chisel it, it will come apart at the mortar, separating in complete individual bricks. Similarly, certain minerals, such as calcite, will break into uniform shapes, again and again. Finally, **luster** is the intensity with which a mineral reflects light—a mineral with glassy luster, for example, is similar to the "shininess" of window glass.

Dangerous Places

Michigan's rich mining history means that there are hundreds of mines and quarries, new and old, that dot the landscape. It is highly unlikely that you will come across an unrestricted mine, but in the event that you do, you must be very careful. As tempting as it may be, you should never enter an old mine as it may be a hundred years old or more. Age and disrepair make these places highly unstable and prone to collapse with human interaction. Equipment and buildings left over at mine sites can also be quite dangerous due to age, rust and weathering and should not be tampered with. Finally, mine dumps, or piles of waste rock left over at mines, are often suggested as mineral-collecting sites in this book. While they are great places to hunt, they can also be dangerous. Many piles are several stories tall and have very loose rock on their surfaces. Climbing or careless digging can easily cause a slide of hundreds of pounds of jagged rock.

 Potentially Hazardous Rocks and Minerals

The vast majority of Michigan's minerals are safe to collect and handle, but there are a few with their own dangers. Potentially hazardous minerals in this book are marked with the symbol shown above; the hazards associated with them are described in the "notes" section on the pages listed below. Always take proper precautions when collecting these minerals.

Galena (page 109)—contains lead; avoid over-handling

Mohawkite (page 163)—contains arsenic; avoid inhalation of dust

Serpentine group (page 205)—some minerals are asbestos; avoid inhalation of dust

Tyuyamunite (page 231)—radioactive; avoid over-handling

Hardness and Streak

There are two important techniques everyone wishing to identify minerals should know: hardness and streak tests. All minerals will yield results in both tests, as will certain rocks, which makes these tests indispensable to collectors.

The measure of how resistant a mineral is to abrasion is called hardness. The most common hardness scale, called the Mohs Hardness Scale, ranges from 1 to 10, with 10 being the hardest. An example of a mineral with a hardness of 1 is talc; it is a chalky mineral that can easily be scratched by your fingernail. An example of a mineral with a hardness of 10 is diamond, which is the hardest naturally occurring substance on earth and will scratch every other mineral. Most minerals, including many of Michigan's, fall somewhere in the range of 2 to 7 on the Mohs Hardness Scale, so learning how to perform a hardness test (also known as a scratch test) is crucial. Common tools used in a hardness test include your fingernail, a copper coin, a piece of glass and a steel pocket knife. There are also hardness kits you can purchase that have a tool of each hardness.

To perform a scratch test, you simply scratch a mineral with a tool of a known hardness—for example, we know a steel knife is a hardness of 5.5. If the mineral is not scratched, you will then move to a tool of greater hardness until the mineral is scratched. If a tool that is 6.5 in hardness scratches your specimen, but a 5.5 did not, you can conclude that your mineral is a 6 in hardness. Two tips to consider: as you will be putting a scratch on the specimen, perform the test on the backside of the piece (or, better yet, on a lower-quality specimen of the same mineral), and start with tools softer in hardness and work your way up. On page 10, you'll find a chart that shows which tools will scratch a mineral of a particular hardness.

The second test every amateur geologist and rock collector should know is streak. When a mineral is crushed or powdered, it will have a distinct color—this color is the same as the streak color. When a mineral is rubbed along a streak plate, it will leave behind a powdery stripe of color, called the streak. This is an important test to perform because sometimes the streak color will differ greatly from the mineral itself. Hematite, for example, is a dark, metallic and gray mineral, yet its streak is a rusty red color. Streak plates are sold in some rock and mineral shops, but if you cannot find one, a simple unglazed piece of porcelain from a hardware store will work. There are only two things you need to remember about streak tests: If the mineral is harder than the streak plate, it will not produce a streak and will instead scratch the plate itself. Secondly, don't bother testing rocks for streak, since they are made up of many different minerals and won't produce a consistent color.

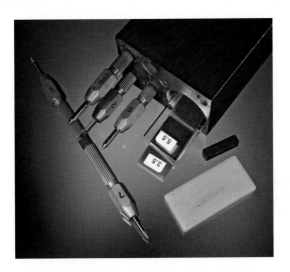

The Mohs Hardness Scale

The Mohs Hardness Scale is the primary measure of mineral hardness. This scale ranges from 1 to 10, from softest to hardest. Ten minerals commonly associated with the scale are listed here. Some common tools used to determine a mineral's hardness are listed here as well. If a mineral is scratched by a tool, you know it is softer than that tool's hardness.

HARDNESS	EXAMPLE MINERAL	TOOL
1	Talc	
2	Gypsum	
2.5		Fingernail
3	Calcite	
3.5		Copper Coin
4	Fluorite	
5	Apatite	
5.5		Glass, Steel Knife
6	Orthoclase	
6.5		Streak Plate
7	Quartz	
8	Topaz	
9	Corundum	
10	Diamond	

For example, if a mineral is scratched by a copper coin but not your fingernail, you can conclude that its hardness is 3, equal to that of calcite. If a mineral is harder than 6.5, or the hardness of a streak plate, it will have no streak and will instead scratch the streak plate itself unless weathered or altered by other, softer minerals.

Quick Identification Guide

Use this quick identification guide to help you determine which rock or mineral you may have found. We've listed the primary color groups and some basic characteristics of the rocks and minerals of Michigan, as well as the page number where you can read more about your possible find. The most common traits for each rock or mineral are listed here, but be aware that your specimen may differ greatly.

	If white or colorless and...	then try...
	Hard, white mineral found as veins within other rocks or minerals	aragonite, page 45
	Bladed crystals that feel very heavy for their size	barite, page 51
	Soft, abundant crystals or blocky masses within other rock	calcite, page 57
	Hard, round specimens with a cauliflower-like outer texture	datolite, page 91
	Small blocky crystals with curved faces and pearl-like luster	dolomite, page 97
	Extremely abundant, hard minerals found mostly in rocks like granite	feldspar, page 101
	Very soft, fibrous crystals that are easily scratched by your fingernail	gypsum, page 121

WHITE OR COLORLESS

(continued)	**If white or colorless and...**	**then try...**
	Transparent cubes that taste like table salt	halite, page 125
	Abundant, soft rock that will fizz in vinegar	limestone, page 145
	Soft, white, lustrous rock that can be scratched by a copper coin	marble, page 155
	Very hard, glassy mineral found as six-sided crystals or as hard, white portions of rock, such as granite	quartz, page 187, 189
	Hard, grainy rock with a grainy appearance	quartzite, page 191
	Tiny white "puff ball" crystals found within cavities in rocks	saponite, page 199
	Fibrous, fan-shaped crystals found within cavities in basalt	zeolites, page 237

WHITE OR COLORLESS

Quick Identification Guide (continued)

If gray and...	then try...
Soft, light mineral found in large masses with gypsum and halite	anhydrite, page 43
Extremely common, dark-colored rock often found on the lakeshore	basalt, page 53
Brass-colored mineral with gray, and sometimes colorfully iridescent, surface tarnish	bornite, page 55
Dark, soft, brittle, metallic mineral found in copper-rich areas	chalcocite, page 63
Very hard, massive material with a waxy surface feel and luster	chert, page 67
Dense, gray rock with a mottled appearance that exhibits lighter spots	diabase, page 93
Dark rock, sometimes with a greenish color, that has large, shiny, coarse-grained crystals within	gabbro, page 107
Extremely heavy, metallic mineral with cubic crystals	galena, page 109
Uncommon, grayish blue rock with fragments of other minerals within	kimberlite, page 137

GRAY

Quick Identification Guide

(continued) **If gray and...** **then try...**

Heavy, gray metallic mineral with spots of quartz within — mohawkite, page 163

Very soft, bluish-gray metallic mineral found as small flakes — molybdenite, page 165

If black and... **then try...**

Glassy, hard crystals or masses found within dark rocks, like gabbro — augite, page 47

Hard, layered pockets of dark material within schist — chloritoid, page 71

Lightweight, soft, shiny, dark rock — coal, page 75

Botryoidal (grape-like) crusts that leave your hands black after handling — cryptomelane, page 87

Botryoidal (grape-like) crusts of a hard, metallic mineral with a reddish streak — hematite, page 127

Dark, shiny mineral found in kimberlite rock — ilmenite, page 131

GRAY

BLACK

(continued)	**If black and...**	**then try...**
	Shiny, metallic mineral which attracts a magnet	magnetite, page 149
	Soft, stubby fan-shaped crystal groupings with a brownish streak	manganite, page 153
	Abundant, soft and thin, very flexible mineral found within rocks	mica, page 159
	Shiny, metallic mineral that forms in fan-shaped crystal groupings with a bluish streak	pyrolusite, page 183
	Soft, dark crystal pockets within limestone	sphalerite, page 213
	Dull, dark mineral found with copper minerals, even growing atop copper itself	tenorite, page 221
	Shiny, metallic veins within quartz or schist	tetrahedrite, page 223
	Long, slender crystals with striated (grooved) faces	tourmaline, page 229

BLACK

Quick Identification Guide (continued)

If blue and...	then try...
Dark blue coatings or crystals occurring with copper and malachite	azurite, page 49
Light blue, blocky or bladed crystals found within pockets in limestone	celestine, page 59
Soft, greenish-blue masses abundant in copper-rich areas	chrysocolla, page 73
Tiny, deep blue, very rare crystals within a blocky, white mineral	kinoite, page 139

BLUE

If yellow and...	then try...
Bright, metallic brass-colored mineral, sometimes with a multi-colored surface tarnish	chalcopyrite, page 65
Very hard, massive material with a waxy surface feel and luster	chert, page 67
Small, fibrous "sprays" of crystals on the surface of rock	grunerite, page 39
Hard, pale brassy-colored metallic mineral with a dark-colored streak	marcasite, page 157

YELLOW

Quick Identification Guide (continued)

(continued) **If yellow and...**	**then try...**
Brassy yellow metallic mineral that forms in cubes	pyrite, page 181
Hexagonal (six-sided) brassy brown metallic mineral that is magnetic	pyrrhotite, page 185
Lemon-yellow pyramid-shaped crystals	sulfur, page 217
Thin, canary-yellow crusts on rock	tyuyamunite, page 231

If green and...	**then try...**
Dark green, very soft mineral often found coating agates or calcite	chlorite, page 69
Grayish-green masses within soft white rock	diopside, page 95
Common, striated (grooved), pea-green crystals found in basalt vesicles (gas bubbles)	epidote, page 99
Soft green crystals or masses within conglomerate rock	fluorite, page 103

Quick Identification Guide (continued)

(continued) **If green and...**	**then try...**
Thin, dark-green coatings on copper or with other copper minerals	malachite, page 151
Yellowish-green masses within rocks, such as gabbro and kimberlite	olivine, page 167
Pale green crystals growing in botryoidal (grape-like) crusts with a fibrous cross section	prehnite, page 175
Tiny, olive-green crystals or beach-worn nuggets	pumpellyite, page 179
Soft, green material with a greasy or silky feel	serpentine, page 205
Extremely soft mineral that is easily scratched by your fingernail	talc, page 219
Very hard mineral found as masses within very coarse rock	topaz, page 227
Masses of hard, intergrown crystals with striated (grooved) faces	tremolite, page 39

GREEN

If brown and...	then try...
Very hard, glassy masses within hard rock	andalusite, page 41
Limestone with forms within it that resemble living things	fossils, page 105
Botryoidal (grape-like) or stalactitic (carrot-shaped) masses that have a fibrous cross section	goethite, page 115
Soft, abundant rock that fizzes in vinegar	limestone, page 145
Rust-colored dusty mineral found as chunky masses	limonite, page 147
Limestone containing six-sided fossil shapes	Petoskey stone, page 169
Highly layered, soft rock found in large beds	shale, page 207
Glassy, light-brown, blocky mineral	siderite, page 209
Very hard, rectangular crystals within hard, layered rock	staurolite, page 215

BROWN

Quick Identification Guide *(continued)*

	If red and...	then try...
	Glassy, ruby-red crystals or masses occurring with copper or tenorite	cuprite, page 89
	Hard, round crystals embedded within rock	garnet, page 111
	Hard, dense, opaque masses that feel waxy to the touch	jasper, page 133
	Abundant, hard, reddish rock that often has bands of color	rhyolite, page 195

RED

	If orange and...	then try...
	Hard, rectangular crystals found within rocks as embedded crystals or as pockets	feldspar, page 101
	Fibrous, soft, crumbly crystals within vesicles (gas bubbles) in rock	laumontite, page 143

ORANGE

	If violet or pink and...	then try...
VIOLET OR PINK	Hard, glassy, purple crystals or masses	amethyst, page 189
	Soft, dark purple masses within very coarse rock	fluorite, page 103
	Rose-colored blocky crystals or masses	rhodochrosite, page 193
	Rare, fibrous nuggets that exhibit "eyes," sometimes with green coloration	thomsonite, page 225

	If metallic and...	then try...
METALLIC	Brassy metallic mineral with a dark gray, often iridescent, colorful tarnish	bornite, page 55
	Dark, soft, brittle metallic mineral found in copper-rich areas	chalcocite, page 63
	Brittle, metallic brass-colored mineral, sometimes with a multi-colored surface tarnish	chalcopyrite, page 65
	Soft, bendable reddish metal, often with black or green tarnish	copper, page 81–85

Quick Identification Guide (continued)

	If metallic and...	then try...
	Extremely heavy metallic mineral with cubic crystals	galena, page 109
	Botryoidal (grape-like) or stalactitic (carrot-shaped) masses that have a fibrous cross section	goethite, page 115
	Soft, bendable, bright yellow metal	gold, page 117
	Reddish metal and gray metal together in the same specimen	halfbreed, page 123
	Black, shiny mineral with a reddish-brown streak color	hematite, page 127,129
	Dark metal that a magnet will stick to	magnetite, page 149
	Soft, stubby fan-shaped crystal groupings with a brownish streak	manganite, page 153
	Hard, pale brassy-colored metallic mineral with a dark-colored streak	marcasite, page 157
	Tiny, thin, delicate brassy-colored needles within geodes	millerite, page 161

METALLIC

Quick Identification Guide (continued)

(continued)	**If metallic and...**	**then try...**
	Heavy, gray metallic mineral with spots of quartz within	mohawkite, page 163
	Very soft, bluish-gray metallic mineral found as small flakes	molybdenite, page 165
	Brassy yellow metallic mineral that forms in cubes	pyrite, page 181
	Shiny, metallic mineral that forms in fan-shaped crystal groupings with a bluish streak	pyrolusite, page 183
	Hexagonal (six-sided) brassy brown metallic mineral that is magnetic	pyrrhotite, page 185
	Soft, bendable, white or grayish metal, often with a dark gray surface tarnish	silver, page 211
	Shiny, metallic veins within quartz or schist	tetrahedrite, page 223

METALLIC

Quick Identification Guide (continued)

	If multi-colored or banded and...	**then try...**
	Hard, translucent, rounded masses containing concentric (bull's-eye) banding within, mostly red and brown	agate, page 29
	Hard, translucent masses with a waxy surface feel	chalcedony, page 61
	Rock made up of many smaller rounded stones or stone fragments	conglomerate, page 77
	Loosely layered rock that resembles rock of another type	gneiss, page 113
	Mottled, or "speckled," hard rock, consisting mostly of light-colored grains with few dark areas	granite, page 147
	Reddish metal and gray metal together in the same specimen	halfbreed, page 123
	Opaque, parallel bands of bright red and dark gray	jasper, page 133
	Soft, fragmented rock that is pink, brown, or tan	Kona dolomite, page 141
	Large, rectangular, sharp-edged crystals within dense, fine-grained rock	porphyry, page 171

MULTI-COLORED OR BANDED

(continued)	**If multi-colored or banded and... then try...**	
	Bright red and dark brown fragments within grainy white rock	pudding stone, page 177
	Layered rock consisting entirely of sand, causing it to feel grainy and rough	sandstone, page 197
	Highly layered, often hard rock that sometimes is very "glittery"	schist, page 201
	Rounded, reddish-brown stones showing yellow or white "spiderweb" cracks	septarian geodes, page 203
	Green, orange and pink stone commonly found on the lakeshore	unakite, page 233

FLUORESCENT

	If fluorescent and...	**then try...**
	Only visible as a yellow or blue fluorescent glow within quartz under short-wave ultraviolet light	powellite, page 173

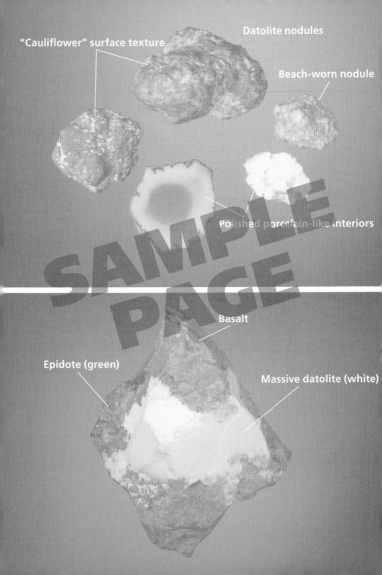

"Cauliflower" surface texture

Datolite nodules

Beach-worn nodule

Polished porcelain-like interiors

Basalt

Epidote (green)

Massive datolite (white)

Sample Page

HARDNESS: 7 **STREAK:** White

Occurrence

ENVIRONMENT: The type of place where this rock or mineral can be found. For the purposes of this book, the primary environments listed include the lakeshores, riverbeds, quarries, which can include gravel pits, road cuts, which are where roads have been cut through hills, and mine dumps, which are the large piles of waste rock left over at the sites of many old mines

WHAT TO LOOK FOR: Common and characteristic traits of the rock or mineral.

SIZE: The general size range of the rock or mineral

COLOR: The general colors the rock or mineral exhibits in its natural state

OCCURRENCE: How easy or difficult this rock or mineral is to find. "Very common" means the material takes virtually no effort to find. "Common" means the material can be found with little effort. "Uncommon" means the material may take a good deal of hunting to find. "Rare" means the material will take great lengths of time and energy to find. "Very rare" means the material is so uncommon that you will be lucky to even find a trace of it.

NOTES: These are additional notes about the rock or mineral that include how to find it, how to identify it, and interesting facts about it.

WHERE TO LOOK: Here you'll find specific places and geographical locations or ranges where you can find the rock or mineral.

Common beach agates

Agate in basalt

Polished agate

Eye agate (polished)

Agate

HARDNESS: 7   **STREAK:** White

Occurrence

ENVIRONMENT: Lakeshore, riverbeds, mine dumps, road cuts

WHAT TO LOOK FOR: Red or brown masses of material that contain concentric (bull's-eye) banding and look and feel waxy

SIZE: Agates are rarely larger than your fist

COLOR: Brown to red, white, yellow, gray to blue; multi-colored banding

OCCURRENCE: Uncommon

NOTES: Agates are perhaps Lake Superior's best known and most widely collected gem. While Minnesota's North Shore is famous for agates, Michigan's Lake Superior shorelines are also great hunting grounds for the popular stones. Agates are a banded variety of chalcedony, which is microcrystalline quartz (quartz crystals too small to see), that form within vesicles (gas bubbles) in basalt. These bands are like the layers of an onion—one on top of another, growing inward. This is known as concentric banding, or banding that resembles a bull's-eye. Nevertheless, agates are very under-researched and while there are many theories, no one knows for certain how the bands form.

As a rule, if a specimen of chalcedony does not have concentric bands, it is not an agate. Agates exhibit all the traits of quartz-based minerals, such as a waxy surface feel and appearance, considerable hardness, and conchoidal fracture (when struck, circular cracks appear). They also share traits with chalcedony, such as translucency and reddish-brown colors caused by iron inclusions, making them easy to identify.

WHERE TO LOOK: Lake Superior's shorelines on the Keweenaw Peninsula are great spots to look, as are the beaches near Grand Marais.

Whole nodules

Broken nodule

Specimens courtesy of Ken Flood

Polished agate

Broken nodules

Specimens courtesy of John Perona

Whole nodules

Agate, Brockway Mountain

HARDNESS: 7 **STREAK:** White

ENVIRONMENT: Lakeshore, riverbeds, road cuts

WHAT TO LOOK FOR: Gray or brown "balls" that contain characteristic agate banding within

SIZE: Brockway Mountain agates are commonly golf ball-sized and smaller

COLOR: Gray, brown, or reddish-brown on the outside; orange, red, yellow, white to tan on the inside

OCCURRENCE: Uncommon

NOTES: The agates found on and near Brockway Mountain, at the tip of the Keweenaw Peninsula, are a unique brand of agate. They generally lack the chlorite coating found on many of the Keweenaw Peninsula's agates and are not the water-worn, waxy specimens found on the lakeshore. Instead, they are found as rough, sometimes almost perfectly round, hard nodules (round mineral clusters), with a brown or gray outer surface color. The interiors of these nodules are distinctly colored in pastel shades of orange, yellow and white, sometimes with very tightly spaced banding. These similarly colored bands lack contrast and can sometimes appear as one solid field of color unless you look very closely.

Brockway Mountain agates can be found over a large area at the tip of the Keweenaw Peninsula. Most are inland and must be dug out of the ground, though there are beaches near Copper Harbor where they can also be found.

WHERE TO LOOK: Look in the area between Eagle River and Copper Harbor on the Keweenaw Peninsula. Look in outcroppings of exposed rock as well as the lakeshore.

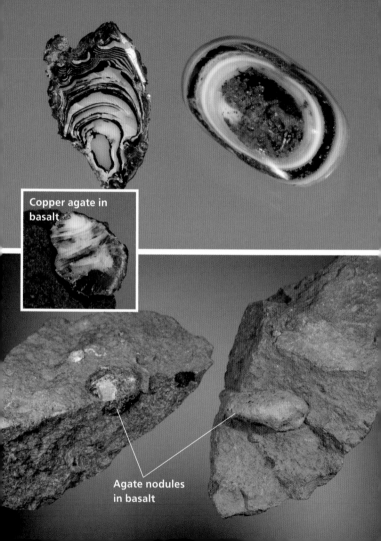

Polished copper-banded agates (specimens about 1 inch long)

Copper agate in basalt

Agate nodules in basalt

Agate, Copper-Banded

HARDNESS: 7 **STREAK:** White

ENVIRONMENT: Mine dumps

Occurrence

WHAT TO LOOK FOR: Dark nodules in basalt that contain characteristic agate banding and copper within

SIZE: Copper-banded agates are very rarely larger than your thumbnail

COLOR: Dark green to gray on the outside; pale orange, tan, white to cream-colored inside with copper-red metal

OCCURRENCE: Very rare

NOTES: Of all the world's rare, valuable, and beautiful agate varieties, none are outdone by the Keweenaw Peninsula's exquisite copper-banded agates. Also called "copper replacement agates," these small, unique gems are often coated with a thin layer of dark-green chlorite that hides the treasures within. A Keweenaw native, John Perona, and his father, Frank Perona, are credited with the discovery of these agates in 1951, found in a copper mine dump. In fact, the mine dumps north of Houghton are the only place in the world where these agates are found, but don't get your hopes up just yet—it takes even the most diligent collector weeks to find just one specimen. The agate nodules (round mineral clusters) must be broken out of solid basalt in order to find high-quality specimens. The leading theory as to how these agates formed is that the agates dehydrated during formation, leaving voids in their banding which then later filled with copper.

WHERE TO LOOK: The copper mine dumps north of Houghton on the Keweenaw Peninsula are the only places to look.

Polished agate

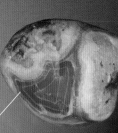

Beach-worn agates

Specimens courtesy of Karen Brzys

Polished agates

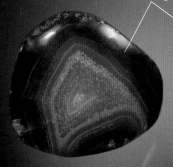

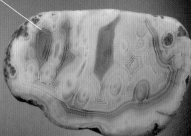

Specimens courtesy of Karen Brzys

Agate, Grand Marais

HARDNESS: 7 **STREAK:** White

Occurrence

ENVIRONMENT: Lakeshore

WHAT TO LOOK FOR: Hard, waxy nodules containing characteristic agate banding

SIZE: Agates are rarely larger than your fist

COLOR: Brown to red, white, gray to blue, black

OCCURRENCE: Rare

NOTES: Grand Marais is a fairly isolated town on the eastern end of the Upper Peninsula, near Lake Superior's beaches. The agates found on those sandy shores, called Grand Marais agates, are unique because they are a rarity there, carried by the immense glaciers of the past ice ages, far removed from the agate-rich regions of Minnesota, Wisconsin and the Keweenaw Peninsula. As if being bulldozed by the glaciers wasn't enough, they have spent centuries in the lake, being rounded and smoothed by the waves and the sand. Unique colors and patterns are often present in Grand Marais agates, suggesting that their original source was a uniquely mineralized deposit.

When hunting for agates anywhere, including the beaches around Grand Marais, remember the hallmarks of quartz-based minerals: a waxy surface feel and appearance, a very high hardness, and conchoidal fracture (when struck, circular cracks appear). In addition, agates are concentrically banded (bull's-eye patterned) chalcedony, so chalcedony's colors and translucency are key.

WHERE TO LOOK: Grand Marais is located on Michigan Highway 77 on the shore of Lake Superior. The sandy beaches extend for miles to the east.

Chlorite-coated agate nodules

Chlorite-coated agate nodules in basalt

Polished agate

Keweenaw agate in basalt (broken open)

Agate, Keweenaw

HARDNESS: 7 **STREAK:** White

ENVIRONMENT: Mine dumps

Occurrence

WHAT TO LOOK FOR: Dark green nodules within basalt that contain characteristic agate banding within

SIZE: Keweenaw agates are generally thumbnail-sized and smaller

COLOR: Green to dark-green on the outside; pale orange, tan, white to cream-colored inside

OCCURRENCE: Rare

NOTES: There are many varieties of agate, depending on where they formed and what minerals they formed alongside. In the Keweenaw Peninsula, agates can be found still embedded within their basalt matrix, or host rock. When whole, their exterior does not exhibit the concentric (bull's-eye) banding we've come to know in agates. Keweenaw agates tend to be coated with a thin, dark-green layer of chlorite. Chlorite is a soft mineral that forms as a lining within the vesicle (gas bubble) before the agate does, thus appearing as the outer surface of the agate nodules (round mineral clusters). Inside these small, round nuggets, however, is the characteristic agate banding. While many agates are multi-colored due to impurities, such as reds and browns caused by iron, Keweenaw agates have been stained less, resulting in pale, light-colored bands. Whites, tans and oranges are common.

Not all chlorite-coated nodules contain agate. Some are solid chlorite, while others contain calcite or quartz.

WHERE TO LOOK: As the name indicates, mine dumps on the Keweenaw Peninsula are the best places to hunt for these agates.

Chalcopyrite

Tremolite

Tremolite

Grunerite

Amphibole group

HARDNESS: 5–6 **STREAK:** White

Occurrence

ENVIRONMENT: Quarries, mine dumps

WHAT TO LOOK FOR: Light-colored, fibrous, elongated crystals, often within or on rock

SIZE: Amphibole minerals can occur as small, pea-sized crystals within rock or as larger masses, sometimes bigger than a basketball

COLOR: White to yellow, pale green, brown to black

OCCURRENCE: Uncommon

NOTES: The amphibole group is a family of fairly hard, rock-building minerals, which are very common ingredients within rocks such as granite. Many amphiboles are dark colored, but not all of them are. Two common exceptions are tremolite and grunerite, two commonly collected amphiboles found on the upper peninsula. While there are other amphibole minerals in Michigan, most others are incorporated within rock and would be difficult to identify. Tremolite is a white to pale-green mineral that forms fibrous masses of crystals, most often within marble (metamorphosed limestone). Its slender, silky crystals are striated (grooved) and are generally easy to identify, though a hardness test will help. Actinolite is a deep-green, iron-rich variety of tremolite—though most Michigan "actinolite" found in shops does not contain enough iron and is actually just green tremolite. Grunerite is generally a golden yellow or brown color, forming fibrous radial "sprays" on the surface of rock, or appearing as small, thin, stubby crystals embedded within rock.

WHERE TO LOOK: The marble quarries near Felch in Dickinson County produce fine specimens of tremolite.

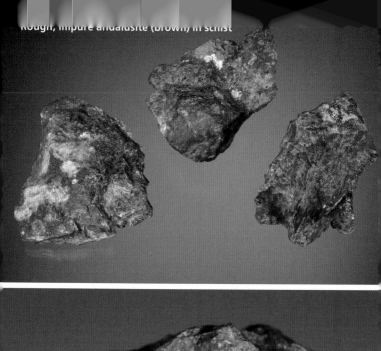

Andalusite

Andalusite

HARDNESS: 7.5 **STREAK:** Colorless

Occurrence

ENVIRONMENT: Mine dumps, roadcuts

WHAT TO LOOK FOR: Very hard, reddish-brown, glassy crystals or masses within rock, namely schist

SIZE: Andalusite crystals are generally golf ball-sized and smaller

COLOR: Brown to reddish-brown, gray

OCCURRENCE: Uncommon

NOTES: Andalusite, an aluminum-rich mineral, is very hard, which is one of its best identifying characteristics, as few other Michigan minerals match its hardness. Like garnets, it is most often a product of metamorphism, which occurs when rocks are subjected to great pressure and heat, and as such it can be found embedded within metamorphic rocks like schist, gneiss and slate (metamorphosed shale). Well-formed crystals are quite rare and occur as thick, stubby, rounded crystals. Andalusite is more commonly found as glassy, brownish-red masses within schists.

Collectors don't often set out to find andalusite as it isn't particularly valuable, except as large, pure masses or fine crystals. However, as it often occurs alongside more collectible minerals derived from metamorphism, such as garnet and staurolite, its presence can serve as an indicator that you're looking in the right area for those minerals. Andalusite generally isn't mistaken for garnet, but staurolite's similar colors and hardness can be confused with andalusite. Staurolite, however, tends to occur in well-formed cross-shaped crystals, whereas andalusite does not.

WHERE TO LOOK: The Lake Michigamme area 40 miles west of Marquette has slates and schists containing andalusite.

Massive rough anhydrite

Massive rough anhydrite

Anhydrite

HARDNESS: 3–3.5 **STREAK:** Off-white

Occurrence

ENVIRONMENT: Quarries, mine dumps

WHAT TO LOOK FOR: Gray, soft mineral that can occur in very large beds between gypsum and halite

SIZE: Anhydrite can be found in nearly any size, from pebbles to boulders

COLOR: White to dark gray, blue

OCCURRENCE: Uncommon

NOTES: Anhydrite's name derives from the Greek words for "without water," named as such because its chemical composition is identical to gypsum, minus its water content. Crystals of anhydrite have been found in Michigan, but they are quite rare and instead anhydrite is found massively. In Southern Michigan, particularly in the Detroit salt mines, anhydrite is found as layers between massive beds of gypsum and halite. These underground gypsum and halite, or "rock salt," beds can be many feet thick, extend for miles, and contain thin layers of dark-colored anhydrite. In other mines or wells (vertical holes into the ground), the anhydrite is found in huge, pure masses, sometimes over 50 feet thick.

Anhydrite is easily altered, or changed, to gypsum when water is introduced to it. This is understandable since their compositions, as mentioned above, only differ in terms of water content. But this also means that anhydrite can be harder to find than one might think. Pure, unaltered anhydrite tends to only be found deep underground, where collectors cannot obtain it.

WHERE TO LOOK: Detroit salt and gypsum mines, in Wayne County, produce anhydrite.

Aragonite crystals

Chalcopyrite crystals

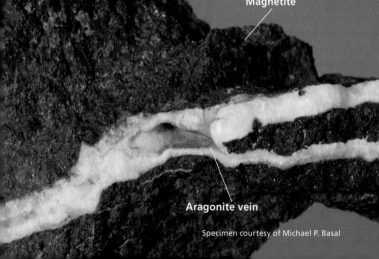

Magnetite

Aragonite vein

Specimen courtesy of Michael P. Basal

Aragonite

HARDNESS: 3.5–4 **STREAK:** White

ENVIRONMENT: Road cuts, mine dumps

Occurrence

WHAT TO LOOK FOR: White crystals, veins, or pockets within rock that are harder than calcite but softer than quartz

SIZE: Individual crystals are generally smaller than a pea; veins can be as long as several feet, depending on the size of the crack they fill

COLOR: Colorless, white to gray, brown

OCCURRENCE: Rare

NOTES: Aragonite is a light-colored mineral that forms small crystals or veins within rock. Its crystals tend to be colorless or white and remain fairly small, which only helps to confuse it with its close mineral cousin, calcite. Calcite and aragonite actually have the same chemical composition, but differ in their crystal structure. It's this difference that creates aragonite's best distinguishing characteristic: it is considerably harder than calcite. However, if a hardness test isn't yielding definite results, assume that your specimen is calcite, as it is much, much easier to find than aragonite. Quartz, another colorless or white crystal common to Michigan, may also confuse collectors, though it is far harder than both calcite and aragonite.

Aragonite is a fairly rare Michigan mineral, primarily found in the Upper Peninsula's iron formations and in some limestone deposits in the Lower Peninsula. Good crystals are very hard to come by and most specimens are veins or massive pieces.

WHERE TO LOOK: The iron-rich area near Ishpeming and Champion has turned up aragonite specimens alongside iron minerals.

Augite (black)

Gabbro

Augite crystal in polished gabbro (about ⅛")

Augite

HARDNESS: 5–6 **STREAK:** Green to gray

Occurrence

ENVIRONMENT: Lakeshore, riverbeds, quarries, road cuts

WHAT TO LOOK FOR: Black, glassy crystals, most often within rock, namely gabbro

SIZE: Augite crystals are generally smaller than your thumbnail

COLOR: Black, dark green, brown

OCCURRENCE: Common as grains embedded in rocks

NOTES: Augite is the most common member of the pyroxene group of minerals, which are dark-colored and important rock-forming minerals. This means that they are most prevalent as grains or crystals that make up rocks, particularly dark volcanic rocks like basalt and gabbro. Augite is common and easily identified within gabbro in particular, appearing as glassy, black masses or crystals, though they are usually quite small and are more easily spotted when a rock specimen is polished. Crystals often have a rectangular or rhombohedral (a shape like a leaning square) cross section and are fairly hard. Collectible varieties are rare in Michigan, though poorly formed bright green masses are sometimes found in kimberlite rock.

Augite is not a widely collected mineral unless it occurs in well-formed loose crystals, which are not found in Michigan. However, due to its frequent occurrence and importance as a rock-building mineral, all collectors should be familiar with augite. Those with the proper rock cutting and polishing equipment may also enjoy the challenge of finding a crystal embedded within gabbro.

WHERE TO LOOK: If you really want to find a sample of augite, look to Lake Superior's shores and find gabbro boulders to break. You'll probably need a magnifying glass as well.

Azurite coating on copper

Azurite (blue)

Specimen courtesy of John Perona

Azurite (blue) with malachite (green) and calcite (white)

Azurite

HARDNESS: 3.5–4 **STREAK:** Light blue

Occurrence

ENVIRONMENT: Mine dumps

WHAT TO LOOK FOR: Deep, rich blue coatings on copper and copper-related minerals

SIZE: Azurite generally occurs as small coatings on copper, but can also occur as thumbnail-sized pockets in copper ores

COLOR: Deep blue to light blue

OCCURRENCE: Rare

NOTES: In other copper-rich regions of the country, particularly Arizona, azurite is a common and widely collected mineral, and it often occurs in large, brilliantly colored crystal points. In Michigan, however, azurite is a rare mineral, and while crystals have been found in several copper mines, all are small and only millimeters in length. Michigan azurite mostly occurs as small, thin, dark-blue coatings on copper or other copper-related minerals. These crusts are of a deep, rich blue color, which distinguishes it from chrysocolla's light bluish-green hues. In addition, chrysocolla is much more common, and is found in nearly any copper deposit.

Azurite contains nearly the same chemical composition as malachite, so as you might imagine, the two minerals very frequently occur together in the same specimen. Confusing the two is very unlikely, as color alone is enough to distinguish one from the other. The only other mineral of very similar color in Michigan is kinoite, though kinoite is extremely rare and occurs only as very minute crystals.

WHERE TO LOOK: There are few known locations for crystals, but deep-blue coatings can be found in most copper mine dumps on the Keweenaw Peninsula.

Lamellar (gill-like) barite crystals

Thicker, glassy barite blades

Barite (Baryte)

HARDNESS: 3–3.5 **STREAK:** White

Occurrence

ENVIRONMENT: Quarries, mine dumps

WHAT TO LOOK FOR: Thin, light-colored crystals that are blade-like and are very heavy for their size

SIZE: Barite crystal clusters are generally smaller than a softball

COLOR: Colorless to gray, yellow to brown

OCCURRENCE: Common

NOTES: Barite, also spelled "baryte," is widespread throughout Michigan, but is hard to find in well-formed crystals. When crystals are present, however, they are fairly easy to identify. Barite commonly forms in thin, tabular (plate-like) crystals that resemble blades that are grouped in lamellar aggregates, which means that the crystal groupings resemble fish gills. In addition, barite has a very high specific gravity, meaning that it always feels very heavy for its size, which is an uncommon trait for a light-colored mineral. Even a small sample can surprise you with its weight. Occasionally, barite can be found as more substantial crystals, occurring as thicker blades ending in fine crystal points. These blades are sometimes stained red with iron, but these are quite rare. Barite also appears as solid veins or pockets within rock.

In the Upper Peninsula, barite is often found in the dumps at iron and copper mines alongside calcite, though rarely as well-formed crystals. Limestone quarries in the Lower Peninsula also produce barite specimens.

WHERE TO LOOK: Iron mine dumps in Marquette County or copper mine dumps in Keweenaw County are great places to look because they have produced some of Michigan's finest specimens.

Rough basalt

Beach-worn basalt

Vesicular basalt

Amygdaloidal basalt

Basalt

HARDNESS: 5–6 **STREAK:** N/A

ENVIRONMENT: All environments

Occurrence

WHAT TO LOOK FOR: Gray to black fine-grained rock, prominent on lakeshores

SIZE: Basalt can be found in any size, from pebbles to cliffs

COLOR: Gray to black, reddish-brown to brown

OCCURRENCE: Very common

NOTES: Basalt is a dark, compact, fine-grained rock that forms as a result of molten rock being spilled onto the earth's surface and cooling very quickly. Unlike very coarse-grained rocks, such as granite, which cooled so slowly that each individual mineral within the rock was allowed time to crystallize to a visible size, basalt cooled too quickly for visible crystals to form. As a result, you need a very strong microscope to see individual crystals or grains of the plagioclase feldspar, augite and olivine that make up most of basalt's mineral composition. This gray, black, or brown rock can easily be found anywhere in the Keweenaw Peninsula and in most of the western end of the Upper Peninsula.

The lava that formed basalt was full of gases that were trapped when it cooled, forming vesicles, or gas bubbles. We call basalt that has many of these gas bubbles "vesicular basalt." Vesicles can then be filled in by other material, creating pockets of minerals called amygdules. Basalt that has many amygdules is called "amygdaloidal basalt" and will have a speckled appearance. Agates and zeolites are two collectible examples of minerals that form within vesicles as amygdules.

WHERE TO LOOK: Basalt can be found on much of the Upper Peninsula's beaches, but primarily on the Keweenaw Peninsula.

Bornite

Specimen courtesy of John Perona

Iridescent, multi-colored coating

Bornite

HARDNESS: 3   **STREAK:** Grayish-black

Occurrence

ENVIRONMENT: Mine dumps

WHAT TO LOOK FOR: Dark metallic mineral, often with an iridescent blue or purple coating

SIZE: Bornite normally occurs massively and can be found in a wide range of sizes, though it most often occurs in palm-sized or smaller pieces

COLOR: Bronze-colored, gray to black or brown, often with deep blue to purple iridescent surface tarnish

OCCURRENCE: Uncommon

NOTES: Bornite is an ore of copper and is often called "peacock ore" due to the multi-colored, iridescent tarnish that forms on its surface. Since the Keweenaw Peninsula is so rich with copper, diligent collectors can find specimens within mine dumps. It is not as common as chalcocite or chalcopyrite, two similar minerals with which it can occur, and telling these three minerals apart isn't always easy. Scratching a specimen will show you its interior color, which is not always the same as the surface color. If it is gray on the outside and gray on the inside, it is most likely chalcocite, another copper ore. If it is a metallic yellow or bronze-colored inside, however, it may be bornite or chalcopyrite, since both minerals have a similar interior color and develop a dark multi-colored tarnish on their exterior surfaces. Hardness and streak tests may help differentiate between the two because chalcopyrite is slightly harder than bornite, and its streak is greenish whereas bornite's is not.

WHERE TO LOOK: Many iron mines in Marquette County produced bornite, as well as much of the Keweenaw Peninsula's copper mines, so looking in mine dumps in those areas would be a great place to start.

Basalt

Copper

Calcite crystals

Calcite-filled cavity in basalt

Calcite rhombohedron

Calcite on limestone

Calcite vein in basalt

Calcite crystal with copper inclusions

Calcite

HARDNESS: 3 **STREAK:** White

ENVIRONMENT: All environments

Occurrence

WHAT TO LOOK FOR: Light-colored crystals, veins, or pockets that are easily scratched with a copper coin

SIZE: Individual crystals tend to be smaller than your palm, but can occasionally be quite large.

COLOR: Colorless to white, yellow to brown

OCCURRENCE: Very common

NOTES: Calcite is one of the world's most common and important minerals, which makes it very easy, and important, to find and identify. It occurs with many different rocks and minerals in virtually every geological environment. Fine calcite crystals are hexagonal (six-sided) prisms, though there are many, many different crystal forms of calcite. It is more commonly found as solid veins or filling vesicles (gas-bubbles) in rocks like basalt, identifiable as soft, white masses. It is generally colorless or white, though impurities can tint crystals various shades of yellow and brown. There are three easy ways to identify calcite. First, its low hardness distinguishes it from similar white or colorless crystals, like aragonite or quartz. Calcite is easily scratched by a copper coin whereas aragonite and quartz are not. Second, calcite effervesces (bubbles) in acid, and even a small drop of vinegar will cause calcite to fizz and dissolve. And thirdly, any specimen of calcite, when carefully broken, will fragment into a perfect rhombohedron (a shape that looks like a leaning cube). This is called cleavage, or the action of a mineral breaking along the planes of its crystalline molecular structure.

WHERE TO LOOK: Look in mine dumps on the Upper Peninsula and pockets within limestone on the Lower Peninsula.

Celestine

HARDNESS: 3–3.5 **STREAK:** White

Occurrence

ENVIRONMENT: Quarries, mine dumps

WHAT TO LOOK FOR: Soft, gray-blue tabular (flat, plate-like) crystals

SIZE: Celestine crystals can become quite large (up to palm-sized), but many crystals are thumbnail-sized or smaller

COLOR: White to gray, blue, bluish-gray

OCCURRENCE: Uncommon

NOTES: Formerly called "celestite," celestine is a unique, collectible mineral found in the southeastern corner of the Lower Peninsula. Many fine celestine crystals are found within vugs (cavities) in limestone, dolostone and sandstone in the quarries near Detroit. These crystals are often bluish, free-standing blades that make for very attractive display specimens. Other crystals are short and stubby, appearing as blocky masses. No matter the form it takes, Michigan's celestine is highly sought-after by collectors, often simply for the fact that Michigan is such a well-known locality for the mineral. Celestine's color and crystal shape should distinguish it from calcite, but if not, a hardness test may help, as celestine is slightly harder. Barite can look similar, but it has a much higher specific gravity (it feels very heavy for its size) whereas celestine does not. Celestine crystals are often color-zoned, or have distinct changes in color. For example, the bottom of the crystal may be white while the top is blue. This can also be a key identifying trait.

WHERE TO LOOK: The mines in Maybee are world-famous sites for celestine, though they are mostly off-limits to collectors.

Chalcedony

Chalcedony nodules

Chalcedony

HARDNESS: 7 **STREAK:** White

Occurrence

ENVIRONMENT: Lakeshore, riverbeds, road cuts, mine dumps

WHAT TO LOOK FOR: Hard, waxy masses of material without a regular or even coloration

SIZE: Chalcedony occurs massively and can range anywhere from pebbles to boulders

COLOR: Brown to red, yellow, white to gray

OCCURRENCE: Very common

NOTES: Chalcedony is a form of microcrystalline quartz, which means that it forms as masses of quartz crystals that are too small to see without a very strong microscope. As a quartz mineral, it exhibits many of the traits of quartz, such as its considerable hardness and conchoidal fracture (when struck, circular cracks appear). It is related to jasper and chert, which are also microcrystalline varieties of quartz, but jasper and chert consist of grains that are highly compacted, preventing light from passing through them. Chalcedony's microcrystals are tiny fibers arranged parallel to each other that easily let light through. The result is one of the primary differences between chert, jasper and chalcedony: chalcedony is translucent, and will "glow" in the sunlight, whereas chert and jasper are opaque.

Agates are a concentrically banded (bull's-eye patterned) variety of chalcedony that form within vesicles (gas bubbles) in basalt as nodules (round mineral clusters). The colors of chalcedony are attributed to impurities; reds and browns, for example, are caused by iron inclusions.

WHERE TO LOOK: Chalcedony is easy to find along Lake Superior's shores as well as nearly anywhere on the Keweenaw Peninsula.

Specimen courtesy of John Perona

Chalcocite vein

Chrysocolla (green)

Chalcocite (metallic gray)

Chalcocite

HARDNESS: 2–2.5 **STREAK:** Shiny black

ENVIRONMENT: Mine dumps

Occurrence

WHAT TO LOOK FOR: Dark, metallic gray, soft mineral occurring in veins within rock

SIZE: Chalcocite normally occurs as masses or veins and can be found in a wide range of sizes; it is commonly found in palm-sized pieces

COLOR: Lead-gray to black

OCCURRENCE: Common

NOTES: Chalcocite is a very economically important ore of copper and has been mined for centuries all over the world, including in Michigan. It is a dark gray metallic mineral that forms in huge veins and deposits in copper-rich areas, though collectors can find it in mine dumps. Identifying chalcocite is easy because few other minerals share its appearance. Chalcopyrite and bornite are similarly colored when tarnished, but a fresh break will reveal these minerals' respective metallic or bronze-yellow colors. In addition, chalcocite is softer than both minerals. It is possible that you could also confuse chalcocite with galena, due to its similar color and hardness, but galena has a cubic structure and cleavage, meaning that when it is carefully broken, galena will break into perfect cubes, but chalcocite will not. Not to mention that galena is a lead mineral that is rare in copper deposits.

When hunting in mine dumps, especially in Ontonagon County, look for dark, metallic chunks or veins in other rock. Chalcocite is often found alongside other copper minerals, especially chrysocolla, which easily forms with any copper minerals.

WHERE TO LOOK: Mine dumps from Ontonagon County all the way up to Keweenaw County have produced chalcocite.

Chalcopyrite

Aragonite

Chalcopyrite crystal on pink dolomite

Chalcopyrite vein

Chalcopyrite

HARDNESS: 3.5–4 **STREAK:** Greenish-black

Occurrence

ENVIRONMENT: Mine dumps, road cuts, river banks

WHAT TO LOOK FOR: Brittle, golden-yellow crystals or veins within rock, sometimes with an iridescent blue surface tarnish

SIZE: Chalcopyrite crystals are generally thumbnail-sized and smaller while massive chalcopyrite can be palm-sized or larger

COLOR: Brassy or golden yellow to brown, sometimes with blue to purple surface tarnish

OCCURRENCE: Common

NOTES: Chalcopyrite is a collectible iron- and copper-bearing mineral that is found in Michigan both as massive pieces as well as crystals. Crystal specimens are more desirable, of course, but they are rare and often exhibit triangular faces or small pyramid-like shapes. Massive specimens are often found as veins within rock, and can sometimes be quite large and as collectible as crystal specimens. Chalcopyrite's bright golden yellow color can be confused with a few other minerals, such as gold and pyrite. Gold is far rarer than chalcopyrite and isn't brittle like chalcopyrite. Pyrite is more common and much harder, and it forms cubic crystals. When chalcopyrite tarnishes, it can develop a dark, multi-colored and iridescent surface coating, similar to bornite's, though bornite is softer and less common.

WHERE TO LOOK: There are dozens of iron mines in Marquette County which makes mine dumps, road cuts and other exposed rock there a great place to start, especially near Ishpeming.

Chert (gray)

Coral fossil in chert

Chert (black) in siltstone (brown)

Banded chert

Chert with
limonite (yellow)

Flint

Chert

HARDNESS: 7 **STREAK:** White

Occurrence

ENVIRONMENT: Lakeshore, riverbeds, quarries

WHAT TO LOOK FOR: Opaque gray or yellow stones that look and feel smooth and waxy

SIZE: Chert occurs massively and can be found in nearly any size

COLOR: Gray to black, yellow to brown

OCCURRENCE: Very common in the Upper Peninsula; uncommon in the Lower Peninsula

NOTES: Chert is often considered a rock rather than a mineral, though it is composed almost entirely of microcrystalline quartz (quartz crystals too small to be seen with the naked eye). The opaque gray or black stones exhibit the traits of other quartz-based rocks and minerals, including its considerable hardness, waxy surface texture and appearance, and conchoidal fracture, meaning that when struck or broken, rounded, half-moon-shaped cracks appear.

Chert is sedimentary and therefore often has layers or bands signifying various stages of formation. It also can contain various fossils, especially coral. In fact, most chert contains some amount of fossilized organic material, mostly algae that cannot be seen with the naked eye. This organic content determines the color of chert—the darker it is, the more organic material it contains. Flint, the name given to black chert, contains a lot of fossilized algae. Yellow, peanut butter-colored chert contains limonite, a variety of iron ore. Chert is sometimes found alongside, or is contained within, other sedimentary rocks, such as siltstone, a fine-grained, dense, brown rock.

WHERE TO LOOK: Lake Superior's shoreline is the best place to look.

Chlorite nodule in basalt

Chamosite chlorite replacement
of almandine garnet

Chlorite replacement of plagioclase
feldspar in basalt

Chlorite group

HARDNESS: 2–2.5 **STREAK:** Colorless

Occurrence

ENVIRONMENT: Mine dumps, road cuts, riverbeds

WHAT TO LOOK FOR: Soft, dark mineral often filling vesicles (gas bubbles) within basalt or coating agates

SIZE: Chlorite crystals are very small, just millimeters in length, but masses can be thumbnail-sized

COLOR: Light green to dark green, gray

OCCURRENCE: Very common

NOTES: The individual members of the chlorite group are very hard to distinguish from one another without laboratory analysis, so you'll normally hear them all referred to simply as "chlorite." There are two varieties in particular, chamosite and clinochlore, which are common in Michigan, but trying to determine which is which without proper equipment is nearly impossible. Both chlorite minerals form in tiny, tabular (flat, plate-like), six-sided crystals that line the inner walls of vesicles (gas bubbles) in basalt, forming a shell. Sometimes the chlorite fills the vesicle completely, forming a solid chlorite nodule (a rounded mineral cluster). Other chlorite samples are found coating agate, calcite, or feldspar, which developed afterwards within the shell of the initial chlorite lining. Chlorite's low hardness, dark green color, and abundance are key identifiers. Chlorite is also a very pervasive mineral and has a way of seeping into other rocks and minerals and changing them. In Marquette County, chamosite replaces almandine garnets to make black, angular balls of chlorite, and on the Keweenaw Peninsula, chlorites do the same to feldspar crystals embedded in basalt.

WHERE TO LOOK: Common in the mine dumps on the Keweenaw.

Schist matrix

Chloritoid (black)

Chloritoid (black) in schist

Chloritoid

HARDNESS: 6.5 **STREAK:** Greenish-gray

Occurrence

ENVIRONMENT: Road cuts

WHAT TO LOOK FOR: Dark-colored plates or masses, generally within schist

SIZE: Chloritoid samples are generally thumbnail-sized

COLOR: Greenish-gray to black

OCCURRENCE: Uncommon

NOTES: Despite its similar name, chloritoid is not related to the chlorite group of minerals. Instead, it is a much harder mineral, often forming as pockets within metamorphic rocks. You'll most likely find it as dark, black or green "knobs" or plates in highly layered schists, since it is a product of metamorphism. In well-formed chloritoid pockets, you can observe its micaceous habit (layers resembling those of mica). Normally, minerals composed of many small, thin layers are quite soft, so chloritoid's unusually high hardness is distinctive. Poorly formed examples could easily be confused with garnets, andalusite, or staurolite, which can all form in the same kind of metamorphic rocks. Garnets, however, are harder and are often red in color. Staurolite forms long, thin crystals that are often twinned (have a second crystal forming through them). And andalusite forms brownish-red glassy crystals and is harder.

Chloritoid is not a well-known mineral and is not widely collected. But it is fairly widespread in small amounts throughout Northern Michigan.

WHERE TO LOOK: The Humboldt area, 25 miles west of Marquette on US-41, and the Lake Michigamme area, 15 miles farther west on US-41, both have schists containing chloritoid.

Chrysocolla in conglomerate

Massive chrysocolla

Chrysocolla on copper

Botryoidal (grape-like) chrysocolla

Chrysocolla

HARDNESS: 2–4 **STREAK:** White to pale blue

Occurrence

ENVIRONMENT: Mine dumps

WHAT TO LOOK FOR: Bright bluish-green colorations on copper or rock

SIZE: Chrysocolla is generally found in masses that are palm-sized or smaller, or as thin coatings on copper

COLOR: Bluish-green to blue, green

OCCURRENCE: Common

NOTES: Chrysocolla is a common and widely collected copper mineral, easily found in any copper-rich area. The bluish-green color is distinctive and normally enough to identify the mineral. It forms easily in copper deposits and is most frequently found as a thin coating of dusty or crumbly material growing on the surface of copper. It also is frequently found as blue pockets within rock, particularly conglomerate, and it can grow to quite a large, collectible size. It is a soft mineral that dehydrates upon exposure to air, making it easily break and fall apart, turning to dust with little provocation. Some collectors keep their specimens in mineral oil to help them retain moisture for as long as possible. Azurite is another blue mineral that forms dusty crusts atop copper, but it is of a much deeper, darker blue color and is much rarer.

"Gem chrysocolla" is sometimes sold in shops or cut for use in jewelry. This is actually a rare variety of quartz that contains impurities of chrysocolla and therefore has the hardness and traits of quartz, not chrysocolla.

WHERE TO LOOK: Any of the copper mine dumps in the Keweenaw Peninsula are likely to have a lot of chrysocolla hidden in the rock pile. Be on the lookout for the blue color.

73

Anthracite

Anthracite

Coal

HARDNESS: >3 **STREAK:** N/A

Occurrence

ENVIRONMENT: Mine dumps, quarries, roadcuts

WHAT TO LOOK FOR: Black, shiny, lightweight rock formed with shale or sandstone

SIZE: Coal occurs massively and can be found in any size, from pebbles to boulders

COLOR: Dark gray to black

OCCURRENCE: Uncommon

NOTES: Some rock collectors *would* like to get a lump of coal for the holidays. Coal is rock formed from the remains of ancient dead plants that never fully decayed. This happened when aquatic plants died and accumulated into layers within acidic waters that contained little oxygen. After being covered and buried by mud and other sediment, the plant material began to undergo various changes. Great pressure above the material compressed it into the first stages of coal development. Peat and lignite are two soft varieties of coal that have only undergone some of the compression necessary to create anthracite, the hardest and most desirable form of coal. In fact, anthracite is the only variety of coal considered to be a true rock, since it has undergone considerable metamorphism (changes due to heat and pressure). With additional metamorphism, anthracite will turn into graphite, a form of pure carbon, and with even further exposure to great heat and pressure, it will turn to diamond. Coal has long been used as a fuel, since its high content of carbon and organic material makes it burn easily.

WHERE TO LOOK: Look around Saginaw Bay, near Saginaw.

Conglomerate

Rounded stones

Specimen courtesy of John Perona

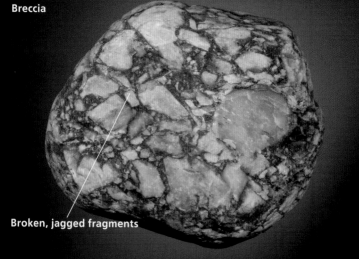

Breccia

Broken, jagged fragments

Conglomerate/Breccia

HARDNESS: Varies **STREAK:** Varies

Occurrence

ENVIRONMENT: Quarries, road cuts, mine dumps

WHAT TO LOOK FOR: Rock that appears to be made of many smaller rocks cemented together

SIZE: Conglomerate and breccia can be found in any size, from pebbles to boulders

COLOR: Varies greatly

OCCURRENCE: Common

NOTES: Conglomerate and breccia are two types of rock made up of many smaller rocks that have become cemented together by another material. The rocks contained within can really be any other kind of rock, depending on how and where the conglomerate or breccia formed.

Conglomerate is made of whole, rounded stones that have been cemented together by a fine-grained rock, particularly sandstone. Frequently, the stones themselves are harder than the cement, and when the conglomerate begins to weather, the stones will stand above the rock's surface. The Upper Peninsula, particularly the Keweenaw Peninsula, has large exposed outcroppings of conglomerate, often containing copper. In fact, many of the Keweenaw Peninsula's copper mines specifically targeted copper-rich conglomerate.

Breccia is made of broken, fragmented stones that have been cemented back together. These particles are often small and jagged, exhibiting sharp edges and appearing as if the original material was violently crushed.

WHERE TO LOOK: The Keweenaw Peninsula has huge lakeside outcroppings of conglomerate.

Sheet copper on shale

Sheet copper

Copper nodule in basalt

Lode copper

Copper

HARDNESS: 2.5–3 **STREAK:** Metallic red

Occurrence

ENVIRONMENT: Mine dumps, lakeshore

WHAT TO LOOK FOR: Flexible, reddish metal, often with black or green surface tarnish

SIZE: Copper can occur in a wide range of sizes, from sheets several feet across to nuggets the size of a pea

COLOR: Copper-red, often with black, green, or red surface tarnish

OCCURRENCE: Common

NOTES: Rock collectors and geologists alike consider Michigan's Keweenaw Peninsula native copper formation the most significant and important copper deposit in the world. It has been mined commercially for over 150 years, with some small-scale mines still operating today, but it was around 3,500 years ago when people first began exploiting the area's copper. Ancient Americans had copper mines of their own, and they would use the metal for making knives, fish hooks and decorations. Michigan's industrial "copper rush" not only removed over 14 billion pounds of copper, but unearthed hundreds of world-class specimens.

There are many ways the soft, malleable (bendable) reddish metal forms. Sheet copper is a common variety, which is found as thin, flat sheets that form between rocks, especially shale. Lode copper, or copper ore, occurs in chunky, branching, intergrown veins that form within rock. Copper also forms as nodules (round mineral clusters) within vesicles (gas bubbles) in rocks, such as basalt.

WHERE TO LOOK: Look all over the entire Keweenaw Peninsula, from Ontonagon to Copper Harbor, especially in mine dumps.

Crystalline copper

"Herringbone" crystal

Prehnite

Copper wire

Specimen courtesy of John Perona

Copper, crystalline

HARDNESS: 2.5–3 **STREAK:** Metallic red

Occurrence

ENVIRONMENT: Mine dumps, lakeshore

WHAT TO LOOK FOR: Flexible, reddish metal, often with a black or green surface tarnish

SIZE: Copper crystals can occur as long as your arm, but most are thumbnail-sized and smaller

COLOR: Copper-red, often with black, green, or red surface tarnish

OCCURRENCE: Rare

NOTES: When people think of crystals, they generally picture the transparent, glassy points of quartz, not branching, tree-like copper crystals. Many amateur collectors are surprised to learn that metals like copper can be found as beautiful, delicate crystals. In fact, copper crystals from the Keweenaw Peninsula grace the collections of mineral museums all over the world.

Copper crystals form in environments where the copper has more room to develop, such as within a large cavity, rather than within cracks or pockets inside rock where other types of copper commonly form. Copper crystals most often form in cubes, though there are many other varieties. Copper "ferns," or tree-like crystals that branch outward, are intricate, and often expensive, examples of copper crystals. Copper wire is a bizarre, rare type of copper crystal that actually consists of very deformed cubic crystals of copper. "Herringbone" crystals, named for their resemblance to fish bones, have a central straight crystal from which others grow like branches.

WHERE TO LOOK: Copper crystals are rarer than most other copper varieties and are only found by the public in copper mine dumps.

"Chisel chips"

Ridges made from hammering action

Specimens courtesy of Ken Flood

Electrolytic copper

Copper, man-made

HARDNESS: 2.5–3 **STREAK:** Metallic red

Occurrence

ENVIRONMENT: Mine dumps

WHAT TO LOOK FOR: Strange, unique shapes of copper found in mine dumps that don't look like natural copper

SIZE: Man-made copper specimens can be palm-sized, but are generally smaller

COLOR: Copper-red, often with black, green, or red surface tarnish

OCCURRENCE: Uncommon

NOTES: As long as people have been exploiting the Keweenaw Peninsula's copper deposits, they have also been inadvertently creating unusual, man-made varieties of copper. These are not always obvious as being man-made. For example, "chisel chips" are very often sold in shops and may resemble sheet copper (page 79), but in fact they were made by a miner's hammer and chisel. The mining operations were generally after copper ore, so when miners encountered pure copper nuggets that were sometimes many feet across and weighed thousands of pounds, they had to cut through them. They did this by hammering their chisel into the nuggets, cutting pieces from them. Therefore, the width of a chisel chip is the same as the width of the chisel used to make it, and the ridges on the surface were made by the repeated hammering. Electrolytic copper is a botryoidal (grape-like) variety made when solutions of water and copper are electrified. It is a by-product of copper-plating processes.

WHERE TO LOOK: Look in mine dumps and around the sites of old copper-processing operations.

Copper replacement of conglomerate boulder (approximate size: 8 inches)

Specimen courtesy of Alex Fagotti

Beach-worn nugget

Float copper

Copper, varieties

HARDNESS: 2.5–3 **STREAK:** Metallic red

Occurrence

ENVIRONMENT: Mine dumps, lakeshore

WHAT TO LOOK FOR: Flexible, reddish metal, often with black or green surface tarnish

SIZE: Copper can occur in a wide range of sizes, from boulders weighing thousands of pounds to pea-sized grains

COLOR: Copper-red, often with black, green, or red surface tarnish

OCCURRENCE: Uncommon

NOTES: Copper is easily identified when it exhibits its classic metallic-red or orange color, but like other metals it can oxidize, or tarnish. A dark gray or black tarnish, called tenorite, commonly forms on most copper specimens, as does a greenish-blue coating, which is actually a mineral called chrysocolla. Scratching through this surface tarnish will reveal the shiny metal hidden below.

Copper often forms in conglomerate, which is a rock made up of many smaller stones that have become cemented together. Occasionally the copper will actually seep into and replace an individual stone within the conglomerate. These are called "copper boulders." Float copper is a very common form and appears as smoothed, rounded nuggets. These were formed when the glaciers of past ice ages bulldozed the land and scraped up lode copper (copper intergrown throughout rock), pulverizing the rock with their immense weight while concentrating it into nuggets. Float copper can also be found on the Keweenaw Peninsula's beaches as reddish, smooth, heavy nuggets.

WHERE TO LOOK: Look all over the entire Keweenaw Peninsula, from Ontonagon to Copper Harbor, especially in mine dumps.

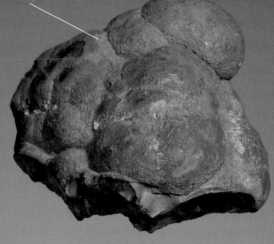

Botryoidal (grape-like) surface structure of psilomelane

Polished psilomelane

Cryptomelane group

HARDNESS: 6–6.5 **STREAK:** Brownish-black

Occurrence

ENVIRONMENT: Mine dumps

WHAT TO LOOK FOR: Black, botryoidal (grape-like) masses that sometimes have a banded cross section and leave your hands black

SIZE: Masses of cryptomelane minerals can be softball-sized and larger

COLOR: Black, gray, brown

OCCURRENCE: Uncommon

NOTES: The word "cryptomelane" comes from the Greek words for "hidden" and "black," in reference to the fact that members of the manganese-based cryptomelane group are all dark-colored and form similarly, thus their identity is "hidden" until laboratory analysis is done. Psilomelane, the most prominent member of this group, isn't a particular mineral at all. Much like how the name "limonite" refers to any unidentified water-bearing iron oxide mineral, the term "psilomelane" describes any manganese oxide with a smooth, botryoidal (grape-like) surface structure—a definition that encompasses a number of minerals. As a collector, it is virtually impossible to tell these minerals apart on your own, especially when most of them exhibit the same structure and hardness. Some psilomelane has a banded cross section which can make for attractive polished specimens. Cryptomelane minerals will often leave a sooty black residue on your hands and streak plate, which is an indicator that the mineral contains the element manganese.

WHERE TO LOOK: The iron mines around Ironwood in Gogebic County have produced wonderful specimens.

Translucent massive cuprite

Red cuprite coating

Chalcotrichite crystals

Tenorite (black)

Cuprite

HARDNESS: 3.5–4 **STREAK:** Brownish-red

Occurrence

ENVIRONMENT: Mine dumps

WHAT TO LOOK FOR: Dark red crystals, coatings, or masses found with copper or tenorite

SIZE: Cuprite crystals are normally quite small (pea-sized or smaller), but crystal aggregates can be palm-sized

COLOR: Deep red to bright red, ruby red, reddish-orange

OCCURRENCE: Uncommon

NOTES: Cuprite is one of the few copper-based minerals found in Michigan's Copper Country that isn't black, green, or blue in color. Instead, cuprite is known for its rich shades of red. Well-formed crystals are cubic, but are very rare in Michigan and cuprite is mostly found as deep red, translucent masses within rock containing copper. Heavily oxidized copper specimens, especially those coming right out of Lake Superior itself, have a thick, opaque, reddish-orange coating of cuprite, frequently contrasted by thick coatings of black tenorite. Cuprite and tenorite are both oxides of copper, so you can expect to see them occurring together very often—even in the same specimen.

Cuprite's color and occurrence with copper are normally enough to identify it. One variety of cuprite that is particularly easy to spot is chalcotrichite, which forms in bright red, lustrous, needle-like crystals. Many of these small crystals form together in a velvety mass within pockets in rock, such as conglomerate, or other copper minerals, like tenorite.

WHERE TO LOOK: As with copper, tenorite and chrysocolla, the copper mine dumps on the Keweenaw Peninsula are your best bet.

Datolite nodules

"Cauliflower" surface texture

Beach-worn nodule

Polished porcelain-like interiors

Basalt

Epidote (green)

Massive datolite (white)

Datolite

HARDNESS: 5–5.5 **STREAK:** White

Occurrence

ENVIRONMENT: Lakeshore, mine dumps

WHAT TO LOOK FOR: Gray or white nodules with a cauliflower-like surface that contain a porcelain-like interior

SIZE: Nodules of datolite are commonly thumbnail-sized, but have rarely been found softball-sized and larger

COLOR: Gray, white, or brown on the outside; white, pink, red to brown on the inside, rarely yellow, green and blue

OCCURRENCE: Uncommon

NOTES: Datolite is one of the Keweenaw Peninsula's unique gems. While datolite is found in many places around the world as small, white, ball-like crystals, Michigan is the only currently known place with compact nodules (round mineral clusters) of the mineral. These lumpy, gray or brown nodules most often have a cauliflower-like texture on their outer surfaces and can be difficult to spot in mine dumps. On the lakeshore, these round masses will be ground-down by the water, smoothing their bumpy shapes and wearing away their dark outer coating, sometimes revealing some of their lighter-colored interiors. Once broken or cut, the real beauty of the specimens can be shown. The hard, porcelain-like core of a datolite nodule is commonly white, gray, or reddish-brown, but much rarer colors, such as mustard-yellow, green and blue are the goal for serious datolite collectors. Datolite nodules form in basalt vesicles (gas bubbles), in which they can sometimes still be found. In shops, you will primarily find datolite specimens cut in half and polished.

WHERE TO LOOK: Lake Superior's shores at the tip of the Keweenaw Peninsula have produced amazing specimens.

"Fuzzy" crystals

Broken surface ("fuzzy" crystals less visible on fresh break)

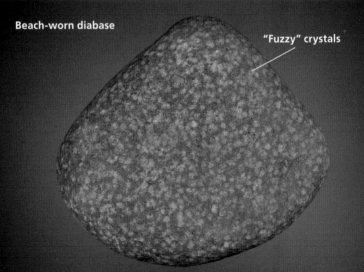

Beach-worn diabase

"Fuzzy" crystals

Diabase

HARDNESS: >5.5   **STREAK:** N/A

Occurrence

ENVIRONMENT: Lakeshore, road cuts, mine dumps

WHAT TO LOOK FOR: Dense, gray rock with "fuzzy" lighter-colored crystals visible

SIZE: Diabase occurs in enormous beds and can be found in any size

COLOR: Gray to black, greenish black, brown

OCCURRENCE: Very common

NOTES: Like basalt, rhyolite and gabbro, diabase is a common volcanic rock found most frequently on Lake Superior's shoreline as beach-worn stones. The dark-colored rock is actually a variety of the much coarser-grained gabbro—a rock that cools deep within the earth and contains many visible crystals. However, diabase cooled much closer to the earth's surface, hardening much faster, and it therefore has far fewer visible crystals than gabbro. These small crystals are a variety of feldspar that is lighter in color than the rest of the rock, giving diabase its mottled or speckled appearance. The feldspar crystals very often appear "fuzzy" or "out-of-focus," meaning that they aren't very well defined. This is because they crystallized first, forcing the other, darker minerals to squeeze in around them, which distorted the feldspar. This is the opposite of what generally happens in rock formation—the dark minerals tend to crystallize first, as in basalt. Specimens with few visible crystals can easily be confused with basalt, and a microscope would then be necessary to distinguish the two.

WHERE TO LOOK: The lakeshore is the easiest place to both find and identify beach-worn diabase boulders.

Massive diopside (green)

Specimen courtesy of Michael P. Basal

Massive diopside (green)

Pyrite

Specimen courtesy of Michael P. Basal

Diopside

HARDNESS: 5–6 **STREAK:** Greenish-gray

Occurrence

ENVIRONMENT: Mine dumps, quarries

WHAT TO LOOK FOR: Gray-green mineral often occurring in metamorphic rocks

SIZE: Diopside generally forms crystals or masses smaller than your palm

COLOR: Light to dark green, grayish green

OCCURRENCE: Rare

NOTES: Diopside, like augite, is in the pyroxene group of minerals, which is a family of important rock-building minerals, meaning that they are very prevalent in the composition of rocks. Unlike augite, however, Michigan's diopside can more often be found as identifiable and collectible masses, rather than just as dark grains within rocks. It is commonly a light to dark grayish-green color, though specimens with inclusions of the element chromium can be a rich, emerald-green color. Diopside can be found embedded within rock, though only two primary types occur in Northern Michigan. In several locations in Dickinson County, diopside is found as green masses embedded within metamorphic rocks like marble and quartzite. It also occurs as grains within kimberlite.

Diopside could be confused with epidote or olivine, but those two minerals are much more common. In addition, epidote has a very distinct yellow-green color and commonly occurs in basalt. Olivine doesn't occur in marbles or quartzites, but is found in kimberlites. For specimens found in that rock type, you'll have to test hardness—olivine is harder.

WHERE TO LOOK: In Dickinson County, the towns of Felch and Iron Mountain both have rocks containing diopside.

Dolomite crystals

Iron-rich dolomite crystals

Coating of clear dolomite crystals

Specimen courtesy of Michael P. Basal

Dolomite

HARDNESS: 3.5–4 **STREAK:** White

Occurrence

ENVIRONMENT: Road cuts, quarries, mine dumps

WHAT TO LOOK FOR: Light-colored groups of slightly curved crystals with a pearly appearance

SIZE: Individual crystals are generally pea-sized or smaller; crystal aggregates can be softball-sized and larger

COLOR: Colorless, white, gray, pink, yellow to brown

OCCURRENCE: Common

NOTES: Dolomite is a common light-colored mineral that is very prevalent in a number of geological environments. It commonly forms in small white or pink blocks that have curved faces, creating what is often referred to as "saddle-shaped" crystals. Its luster is most often pearly and is a good characteristic to note when trying to identify it. Calcite is the only mineral you're likely to confuse it with, though dolomite is considerably harder than calcite. While often found in the Upper Peninsula's iron mines, dolomite is more at home in the Lower Peninsula's limestone deposits where fine crystals are unearthed in quarries.

Since dolomite is so abundant in limestone deposits, it's only natural that some limestone becomes saturated with dolomite. This magnesium-rich limestone was long called "dolomite," though is now referred to as "dolostone" or "dolomite rock" to avoid confusion with the mineral. "Kona Dolomite" (page 141) is the name given to dolomite-rich marble from the Kona Hills near Marquette (on the Upper Peninsula) and often contains fossil algae.

WHERE TO LOOK: The Menominee Iron Range in Dickinson County, near Iron Mountain, has produced excellent crystals.

Epidote crystals

Basalt

Epidote-lined vesicle

Quartz (white)

Green epidote coatings

Epidote crystals

Epidote

HARDNESS: 6–7 **STREAK:** Colorless to gray

ENVIRONMENT: Mine dumps, lakeshore, road cuts

Occurrence

WHAT TO LOOK FOR: Yellow-green coatings or crystals on or in basalt

SIZE: Individual crystals are just millimeters long, but crystal aggregates can be several inches in size

COLOR: Yellowish-green color is distinctive; can also be brownish-green

OCCURRENCE: Common

NOTES: It's easy to find and identify epidote, as it frequently occurs in or on basalt, primarily in mine dumps and on the lakeshore. Epidote forms small, but elongated, crystals that are tabular (plate-like), quite hard and have striated (grooved) faces. Epidote crystals often appear in groups alongside quartz within vesicles (gas bubbles) in basalt. It's epidote's color, however, that makes it so easy to identify. It is most commonly found in a distinct yellow-green color that, once you learn it, is very easy to spot. Even very dark, deeply colored specimens exhibit this color. It is also common to see very thin sheets of this pea-green color on the surface of basalt—these epidote coatings form when a small fissure or crack in the rock fills with a thin layer of epidote. It is possible that you could confuse epidote with olivine, but olivine is harder and doesn't form in vesicles. A green and orange rock called unakite is common on the shore and exhibits epidote's trademark color—that's because unakite is a variety of granite very rich in epidote.

WHERE TO LOOK: Epidote is easily found in any copper mine dump on the Keweenaw Peninsula, as well on the lakeshore.

Microcline-lined vesicles (gas bubbles) in basalt

Massive feldspar

Feldspar crystals in vesicle

Feldspar (orange) in pegmatite

Feldspar porphyry in basalt

Feldspar (pink) in granite

Feldspar group

HARDNESS: 6–6.5 **STREAK:** White

Occurrence

ENVIRONMENT: All environments

WHAT TO LOOK FOR: Very abundant orange crystals or pockets within rocks

SIZE: Feldspar crystals can reach several inches in length, but are generally found embedded within rock and only an inch or less in length and less than a quarter of an inch wide

COLOR: White to cream-colored, flesh-colored, orange, brown

OCCURRENCE: Very common

NOTES: The feldspar group is the most common mineral group on earth, making up nearly 60% of the earth's crust. The term "feldspar" encompasses over a dozen different minerals, several of which are common in Michigan. Within the feldspar group, there are two main subgroups: the potassium feldspars, which include two common varieties—orthoclase and microcline—and the plagioclase feldspars, which include the common minerals albite and anorthite. The vast majority of all feldspars, however, are contained within volcanic rocks, particularly granite, and are only visible as grains or poorly formed crystals. However, there are some collectible specimens to be had—look for hard, opaque, white to pink or orange crystals in pockets, vesicles (gas bubbles), or masses within rocks. Pegmatite and porphyry are particularly good examples of feldspar embedded in rock.

Feldspars are relatively easy to identify as a "feldspar," but determining exactly which mineral you have can be very difficult. Most are simply given a general label, such as "plagioclase," which could actually be one of many minerals.

WHERE TO LOOK: Feldspar is easily found all over the Keweenaw Peninsula.

Fluorite (green) in conglomerate

Specimen courtesy of John Perona
Inset specimen courtesy of Shawn M. Carlson

Fluorite (purple) in pegmatite

Mica crystals

Feldspar

Quartz

Fluorite (purple)

Specimen courtesy of Michael P. Basal

Fluorite

HARDNESS: 4 **STREAK:** White

Occurrence

ENVIRONMENT: Road cuts, quarries, mine dumps

WHAT TO LOOK FOR: Crystals that are softer than quartz, but harder than calcite, growing in masses of rock

SIZE: Fluorite crystals or pockets are generally smaller than a golf ball

COLOR: White, purple, green, yellow to brown

OCCURRENCE: Uncommon

NOTES: Fluorite is a popular collectible mineral that, unfortunately, is quite scarce in Michigan. Its glassy crystals, when well formed, are cubic or octahedral (eight-faced), though only the finest fluorite occurrences, primarily those in the limestone quarries south of Detroit (Lower Peninsula), have produced excellent crystals. Elsewhere, especially in the Upper Peninsula, fluorite is most often found as pockets within rock, and is quite hard to find. Fluorite's crystal shape and colors are its best distinguishing characteristics, though a colorless or white massive specimen can be difficult to identify.

The enormous beds of conglomerate rock found in the Keweenaw Peninsula, particularly near Copper Harbor, are home to beautiful green veins of masses of fluorite and, rarely, the occasional crystal embedded in calcite. The Black River Pegmatite Formation in Marquette County (Upper Peninsula) is a readily available source of poorly formed, purple fluorite masses embedded within the quartz, feldspar, and mica-rich rock.

WHERE TO LOOK: Limestone quarries near Detroit produce excellent crystals alongside celestine.

Petrified wood

Beach-worn coral

Specimen courtesy of Alex Fagotti

Horn coral fossils

Fossils

HARDNESS: Varies **STREAK:** Varies

Occurrence

ENVIRONMENT: Lakeshore, riverbeds, road cuts, quarries

WHAT TO LOOK FOR: Rocks with the appearance of living plants or animals

SIZE: Fossils are generally softball-sized or smaller, though it depends on the species

COLOR: Varies greatly; generally gray, yellow to brown

OCCURRENCE: Uncommon

NOTES: Fossils are the result of ancient plants and animals turning to rock over the course of millions of years. Fossils form when the remains of an organism become covered in sediment, particularly at the bottom of a body of water or within mud or clay that prevents the creature from decaying normally. Minerals then begin to slowly seep into the cells of the once-living organism and replace its tissue with rock. Fossils are commonly found in limestone, and with diligence, you'll likely find some in Michigan's huge limestone deposits.

The lakeshores in both the Upper and Lower Peninsulas are good places to look for beach-worn fossils, particularly corals. These stones appear as common limestone, but exhibit strange segmented, columnar structures. Horn coral, a particular species of coral fossil frequently found in Michigan, can be found as cone-shaped stones with a honeycomb-like cross section. On the Lower Peninsula, Petoskey Stone (page 169) is found on Lake Michigan's shores and is a particularly large, six-sided coral.

WHERE TO LOOK: Horn coral is frequently found on the sandy beaches near Grand Marais. In the Lower Peninsula, any limestone deposit—and there are many—are good places to look, especially near Charlevoix and Alpena. **105**

Beach-worn gabbro

Crystals in gabbro

Rough gabbro

Gabbro

HARDNESS: >5.5   **STREAK:** N/A

ENVIRONMENT: Lakeshore, road cuts

Occurrence

WHAT TO LOOK FOR: Dark, coarse-grained rock containing long, reflective crystals

SIZE: Gabbro occurs in immense formations and can be found in any size, from pebbles to boulders

COLOR: Black to gray, greenish

OCCURRENCE: Common

NOTES: Like granite, gabbro formed from molten rock that cooled deep within the earth. Because it cooled at such great depths, it solidified very slowly, which allowed the various minerals within it enough time to crystallize to a visible size, unlike basalt, which cooled very quickly on the earth's surface and has no visible crystals. Gabbro, therefore, is a coarse-grained, dark rock with many individual minerals easily observed. Many of the small, poorly formed crystals are quite reflective, which can give gabbro a "sparkly" appearance in the sunlight. It is darker than any granite, which is the only rock type you could confuse with gabbro.

Gabbro contains much of the same mineral composition as basalt and diabase, and includes plagioclase feldspars, augite, and olivine. Some of the feldspars are so well developed that large rectangular, striated (grooved) crystals are easily visible and reflect light brightly. Polishing a specimen of gabbro can reveal these as well as well-formed augite crystals.

WHERE TO LOOK: Lake Superior's shores are the easiest place to find and identify weathered boulders of gabbro.

Galena (metallic black) in quartz

Chalcopyrite

Cubic structure of galena

⚠ Galena

HARDNESS: 2.5   **STREAK:** Lead gray

Occurrence

ENVIRONMENT: Quarries, mine dumps

WHAT TO LOOK FOR: A dark, metallic, very heavy mineral, often with cubic crystals

SIZE: Masses of galena are generally found in softball-sized specimens or smaller, but they can occur in much larger formations

COLOR: Dark lead-gray

OCCURRENCE: Uncommon

NOTES: Galena is the world's primary source of the element lead, a very soft, extremely dense metal notorious for its adverse health effects. It has a very high specific gravity (it feels very heavy for its size) and occurs in dark gray metallic masses or, when very well formed, cubic crystals. Even poorly formed specimens will often exhibit cubic structures on the mineral's surface, which is a key identifying trait. When cubic shapes aren't present, you'll have to rely on other tests. Its hardness and streak distinguish it from other similar minerals except for chalcocite. In this case, carefully breaking a piece of galena will yield perfect cubes, whereas chalcocite will not. This is called cleavage, or the action of a mineral breaking along the planes of its crystalline molecular structure.

Galena can be found in small quantities in both peninsulas, and is often associated with sedimentary rocks such as limestone and shale. Despite its lead content, galena is relatively safe to collect and handle, though you will always want to wash your hands well and be careful not to inhale any dust produced when working with it.

WHERE TO LOOK: Limestones in Huron County, in the eastern Lower Peninsula, have been found to contain massive galena.

Schist specimen courtesy of Michael P. Basal

Almandine

Garnet in chlorite schist

Loose specimens courtesy of John Perona
Inset specimen courtesy of Shawn M. Carlson

Pyrope in kimberlite

Almandine garnets replaced by chlorite

Garnet group

HARDNESS: 6.5–7.5 **STREAK:** Colorless

Occurrence

ENVIRONMENT: Quarries, mine dumps, road cuts

WHAT TO LOOK FOR: Very hard, round crystals most often embedded with rock, namely schists and kimberlite

SIZE: Garnets are rarely fist-sized with most being thumbnail-sized and smaller

COLOR: Deep red, brown to black

OCCURRENCE: Uncommon

NOTES: The garnet group consists of several closely related minerals that tend to crystallize in round, multi-sided ball-like crystals. Northern Michigan only has a few varieties of this large group, with the primary collectible types being almandine, spessartine and pyrope. When rocks are changed by heat and pressure, called metamorphism, it often concentrates certain minerals into pockets. This is how most garnets are formed, and many are found embedded within schist, particularly chlorite-based schists. Garnets are very hard, and most of Michigan's varieties are colored shades of red or brown to black. Their colors and hardness are normally enough to identify them, as is their occurrence in schists.

Bright red, gem-quality pyrope garnet is found as small round pockets within kimberlite and can make for very attractive specimens. In Marquette County, some almandine garnets that had formed within chlorite-based schist have since been replaced by the chlorite itself, resulting in dark-colored chlorite pseudomorphs (a mineral with the outward appearance of a completely different mineral).

WHERE TO LOOK: Marquette County has many garnet locations; the pseudomorphs are found near Michigamme on US-41.

Granitic gneiss

Granitic gneiss

Gneiss

HARDNESS: N/A **STREAK:** N/A

Occurrence

ENVIRONMENT: Lakeshore, road cuts

WHAT TO LOOK FOR: Coarse banded or layered rock, often with the appearance of a different rock

SIZE: Gneiss can be found in any size, from pebbles to boulders, though most beach-worn specimens are softball-sized or smaller

COLOR: Varies greatly; often gray, white, black

OCCURRENCE: Common

NOTES: When a rock is subjected to great heat and pressure it can change completely—this is called metamorphism. What commonly happens is that the various minerals within the rock begin to order themselves into layers, or bands. The more layers a rock has and the tighter the spacing between layers, the more metamorphism it has undergone. Gneiss (pronounced "nice") is an example of rock that has only been partially metamorphosed, and while it generally contains bands of separated minerals, it still retains some of the original rock's appearance. Schist, another layered metamorphic rock, has undergone much more of a change and appears in compact, tightly spaced layers. By traditional definitions, gneiss is a rock that has less than half of its minerals ordered into bands, therefore still resembling the original rock. Schist is a rock that has more than half of its minerals arranged into layers and no longer looks like the original rock. Granitic gneiss is common on the lakeshore, and was originally granite, so it appears as a white or pink rock with black bands.

WHERE TO LOOK: The lakeshore is the easiest place to spot gneiss.

Fibrous cross section

Stalactitic masses

Radiating fibers

Polished goethite

Stalactitic mass

"Golden goethite"

Goethite-rich quartz

Goethite

HARDNESS: 5–5.5 **STREAK:** Brown to yellow

Occurrence

ENVIRONMENT: Quarries, mine dumps

WHAT TO LOOK FOR: Metallic, dark mineral often with stalactitic, or icicle-like, formations with a fibrous cross section

SIZE: Goethite masses can be very large, but individual crystals tend to be palm-sized or smaller

COLOR: Black to brown, yellow to yellow-brown

OCCURRENCE: Common in the Upper Peninsula; uncommon in the Lower Peninsula

NOTES: Like hematite, goethite (pronounced "ger-tite") is a very common and important iron ore, widely mined all over the world. But that's not the only trait it shares with hematite—both minerals are brownish-black, metallic, can form botryoidal (grape-like) crusts, and have fibrous cross sections. So how do you distinguish one from the other? The foremost, and most easily testable difference, is that their streak colors are very distinct—hematite's is a reddish-brown while goethite's is a yellowish-brown. Goethite also tends to frequently take a stalactitic (icicle-like) form where hematite does not. Finally, when oxidized or weathered, goethite turns shades of brown to yellow, whereas hematite generally turns shades of red.

The Upper Peninsula's iron districts near Marquette, Ironwood and Iron Mountain are rich with goethite. Some very fine-quality fibrous specimens are of a bright, lustrous honey-yellow color. These are known as "golden goethite," and specimens are quite rare and valuable.

WHERE TO LOOK: Specimens from the Marquette Iron Range, near Marquette and Ishpeming, are among the best to be had.

Gold vein in quartz (specimen size: about ½ inch)

Gold

Quartz

Gold-bearing ore

Gold

HARDNESS: 2.5 **STREAK:** Golden yellow

Occurrence

ENVIRONMENT: River beds, quarries

WHAT TO LOOK FOR: Bright yellow metal, often embedded within quartz

SIZE: Most gold veins and nuggets are pea-sized or smaller, often smaller than a grain of sand

COLOR: Metallic yellow

OCCURRENCE: Very rare

NOTES: Gold—the very name makes any collector's ears perk up. The truth is that most states have gold to collect, you just need to know where to look for it. There are two primary ways which gold is found: as veins within rock or as placer nuggets. Gold veins are irregularly shaped masses that are encased within a rock or mineral, such as quartz. Placer gold is found at the bottom of rivers as small, water-worn pieces, most often smaller than a grain of sand. The term placer (pronounced "plasser") refers to concentrations of dense, heavy grains of material (especially gold) that accumulate at the bottoms of rivers. "Gold ore" is often sold in shops to tourists, and while it may contain tiny inclusions of gold, it is mostly quartz with some pyrite.

Gold is extremely easy to identify because of its low hardness, great malleability (it bends without breaking), and the fact that it never tarnishes, therefore always appearing in its famous metallic-yellow color, all of which are key traits. Pyrite, commonly called "fool's gold," is a very similar color, but is much harder, quite brittle, and has a greenish streak.

WHERE TO LOOK: The Michigan Gold District, just minutes north of Ishpeming, has several mines that produced gold.

Beach-worn quartz-rich (white) granite

Beach-worn feldspar-rich (pink) granite

Granite

HARDNESS: N/A **STREAK:** N/A

Occurrence

ENVIRONMENT: Lakeshore

WHAT TO LOOK FOR: Coarse-grained rock containing many different minerals, each easily seen with the naked eye

SIZE: Beach-worn granite is generally smaller than a basketball

COLOR: Varies greatly; primarily white, pink, gray, with black spots

OCCURRENCE: Common

NOTES: Granite is a common beach rock that is easy to spot and identify. It formed deep within the earth, which made it cool very slowly. When rocks take long periods of time to cool, it allows the individual minerals within them to crystallize to a large, visible size. This is unlike rocks such as basalt, which cooled very quickly on the earth's surface and have no visible individual minerals. Therefore, granite's larger mineral grains give it a mottled, speckled appearance. Granite consists primarily of quartz and feldspars, which contribute to its lighter hues, such as whites and pinks, but it also contains micas and pyroxenes, which make up its darker spots. It's this multi-colored, coarsely grained appearance that makes granite easy to identify.

Granite underlies much of the earth's surface, but doesn't often rise up to where we can see it. The Canadian Shield, an enormous granite formation covering most of Canada, is the nearest place where granite is found on the surface. So how did it get to Michigan's shores? The answer is the glaciers of past ice ages. They scraped up material from the Canadian Shield and brought it farther south.

WHERE TO LOOK: Since most of Michigan's granite is from Canada, the lakeshores are the best place to look.

Fibrous gypsum (selenite)

Massive gypsum

Fine-quality selenite crystals

Gypsum

HARDNESS: 1.5–2 **STREAK:** White

Occurrence

ENVIRONMENT: Quarries, mine dumps, road cuts

WHAT TO LOOK FOR: Light-colored crystals or masses that are easily scratched by your fingernail

SIZE: Masses of gypsum can be many feet thick; individual crystals tend to be palm-sized or smaller

COLOR: Colorless, white, gray, yellow to brown or reddish

OCCURRENCE: Very common in the Lower Peninsula; uncommon in the Upper Peninsula

NOTES: Used as the primary ingredient in plaster and drywall for decades, gypsum is a common mineral most frequently found in sedimentary rock deposits, especially limestone. It is extremely soft and is easily scratched by your fingernail. Gypsum is generally white in color, but it often is tinted in shades of yellow, brown, or red depending on its impurities and in what rock it formed. Its very low hardness, combined with its light colors and sedimentary habit, make gypsum very easy to identify.

By far, the most collectible variety of gypsum is selenite, which occurs in fibrous or glassy crystals. Most gypsum, including selenite is found primarily in the Lower Peninsula in limestone-rich areas, particularly quarries. In Northern Michigan, gypsum is really only found in the sedimentary rocks that cover the eastern end of the Upper Peninsula, in much the same environments as in the Lower Peninsula.

WHERE TO LOOK: Large gypsum beds are quarried in Iosco and Kent Counties in Southern Michigan. Several iron mines in Iron County also produce small amounts of gypsum crystals.

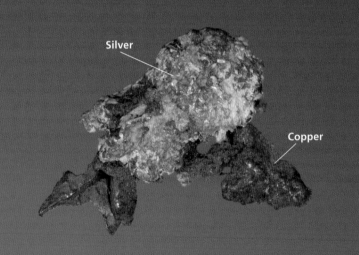

Halfbreed

HARDNESS: 2.5–3 **STREAK:** Varies

Occurrence

ENVIRONMENT: Mine dumps

WHAT TO LOOK FOR: Silver and copper-colored metals together in one specimen, often with a black or gray surface tarnish

SIZE: Halfbreeds are generally no larger than a golf ball

COLOR: Copper red and silver gray, often with gray to black surface tarnish

OCCURRENCE: Rare

NOTES: Halfbreeds are a rare, highly desirable collectible found in the Keweenaw Peninsula's many copper mine dumps. The term "halfbreed" is more of a collector's name and less of a scientific label, but it refers to natural formations of copper and silver together in the same specimen. Technically a rock, since it contains more than one mineral, halfbreeds form because copper and silver occur in the same mineral environments, sometimes atop one another. These prized oddities can come in the form of crystals as well as nuggets or veins.

Identifying a halfbreed is as easy as identifying copper and silver individually. Both metals can tarnish to a dark gray color, but a scratch will reveal their true colors beneath. Their hardness and malleability (they bend rather than break) are also key. Since copper is more common, it is often a larger piece of copper that has smaller pieces of silver upon it, though the opposite can be true as well. When hunting for copper, inspect any specimen you find carefully for silver. Particularly fine, large specimens of halfbreeds are found in shops for very high prices.

WHERE TO LOOK: Keweenaw Peninsula copper mine dumps are just about the only places you'll find halfbreeds.

Rock salt

Halite crystals

Halite cube

Cluster of cubic halite crystals

Halite

HARDNESS: 2.5 **STREAK:** White

Occurrence

ENVIRONMENT: Mine dumps, quarries

WHAT TO LOOK FOR: Clear, cubic crystals that dissolve easily in water and have a salty taste

SIZE: Halite cubes can be as large as your fist, sometimes larger, but generally are thumbnail-sized; massive halite, or "rock salt," can be any size

COLOR: Colorless to white, rarely pink

OCCURRENCE: Common

NOTES: Whether you know it or not, you've no doubt already seen, purchased and even tasted halite crystals. That's because table salt is actually the mineral halite, and many people don't realize that salt must be mined before it can flavor your food. Halite forms in perfectly transparent or white cubes that are hard to mistake for anything else. They dissolve easily in water, are very soft, and have perfect cubic cleavage. This means that when you carefully strike a piece of halite, it will simply break into smaller cubes. If you closely examine a grain of table salt, you'll see that it, too, has a cubic structure. Halite, of course, also has a salty taste, which is a key identifying trait. Avoid tasting natural halite too much, however, since it can sometimes contain harmful impurities as well as destroy the specimen.

Massive, irregularly shaped pieces of halite are sometimes incorrectly called "rock salt." That term actually applies to a rock containing over 95% halite and impurities of anhydrite, gypsum and dolomite.

WHERE TO LOOK: The salt mines near Detroit produce many fine cubic specimens, which sometimes grow on old mining tools.

Varieties of hematite

Fibrous hematite

Granular hematite

Jasper colored red hematite

Red surface oxidation

Botryoidal (grape-like) masses

Hematite

HARDNESS: 5–6 **STREAK:** Brownish-red

Occurrence

ENVIRONMENT: Mine dumps, road cuts, riverbeds, lakeshore

WHAT TO LOOK FOR: Dark gray metallic mineral with a fibrous cross section and red surface oxidation

SIZE: Hematite can occur in very large deposits with some pieces as large as boulders, though most specimens are palm-sized and smaller

COLOR: Steel-gray to black, brownish-red, with red oxidation

OCCURRENCE: Common

NOTES: Hematite is the most common iron-based mineral in the world and has been mined for decades in Michigan. Upper Michigan has three primary iron-rich areas where hematite, and its cousin goethite, are mined and can easily be found. These are the Gogebic Range, near Ironwood, the Menominee Range, near Iron Mountain, and the Marquette Range, near Marquette, and all are good areas to hunt for hematite and other iron-related minerals. This important ore greatly resembles goethite in that it forms most often as dark brown or black botryoidal (grape-like) masses with a fibrous cross section. Though these similarities can make distinguishing the two minerals difficult, there is one primary difference: streak color. Hematite's streak is a distinctive reddish-brown whereas goethite's is a yellow-brown; when oxidized, both minerals remain these respective colors. Hematite is also quite common as reddish stain within other minerals. Agates, quartz and jaspers get their reddish hues from tiny grains of hematite that oxidized within them.

WHERE TO LOOK: Dumps in the three iron ranges listed above, as well as the area near Ishpeming, are the best places to look.

Quartz

Hematite schist

Specular hematite

Specular hematite

Quartz

Micaceous hematite

Hematite, varieties

HARDNESS: 5–6 **STREAK:** Brownish-red

Occurrence

ENVIRONMENT: Mine dumps, road cuts, riverbeds

WHAT TO LOOK FOR: Reflective, "glittery" specimens of hematite that appear to be layered

SIZE: Hematite can occur in very large deposits with some pieces being as large as boulders, though most specimens are palm-sized and smaller

COLOR: Steel-gray to black, brownish-red, with red oxidation

OCCURRENCE: Common

NOTES: Hematite can develop in enormous formations, including vast beds and layers alongside jasper, rather than just as small veins or pockets. This explains why it is such a commonly mined iron ore. But sometimes these huge hematite deposits are metamorphosed, or subjected to heat and pressure, which can completely change their appearance. This is the cause of the huge amounts of specular and micaceous hematite found in Michigan.

Specular hematite, sometimes called "specularite," is a highly reflective, "glittery" form that consists of many small flakes of hematite. It is easily identified because of its high luster, grainy or flaky appearance, and many specimens will leave "glitter" on your hands after handling them. Micaceous hematite, or hematite that forms in thin, stacked sheets resembling mica, is the same thing as specular hematite, but is more metamorphosed. Both types can actually be classified as hematite schists, or highly layered hematite formations.

WHERE TO LOOK: Specular hematite can be found near Champion and Republic, even in roadside ditches.

Magnesium-rich ilmenite

Kimberlite matrix

Specimen courtesy of Shawn M. Carlson

Rough ilmenite

Specimen courtesy of Shawn M. Carlson

Ilmenite

HARDNESS: 5–6 **STREAK:** Brownish-black

ENVIRONMENT: Mine dumps, quarries

WHAT TO LOOK FOR: Metallic black, brittle mineral pockets within rock, especially kimberlite

SIZE: Ilmenite crystals are generally smaller than a golf ball, with most being thumbnail-sized or smaller

COLOR: Black, brownish-black

OCCURRENCE: Common

NOTES: This important ore of titanium is rather common, but not in a very collectible sense. It is a frequent constituent of gabbro, where it forms as embedded grains that differ little from the appearance of the rest of the rock. When it weathers out of rock, it becomes one of the minerals found as sand grains on black sand beaches. But in neither of these two occurrences is ilmenite easy to identify, and therefore can't be considered very collectible. But a magnesium-rich variety of ilmenite, called picroilmenite, is found within all Northern Michigan kimberlite formations as dark, lustrous masses and is more easily identified as well as more collectible. These ilmenites generally don't exhibit any crystal shape and are merely rounded masses within the rock.

Loose ilmenite, especially those metallic specimens found within sand, can resemble hematite or magnetite. However, hematite's streak is reddish-brown, whereas ilmenite's is dark brown. Iron-rich ilmenite can be magnetic, but only weakly so, sticking loosely to a magnet. This differs from magnetite's strong magnetism, which causes it to bond tightly with a magnet.

WHERE TO LOOK: The Lake Ellen Kimberlite Formation is about 15 miles northeast of Crystal Falls, but can be hard to find.

Beach-worn jasper

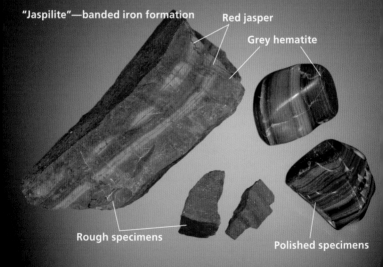

"Jaspilite"—banded iron formation

Red jasper

Grey hematite

Rough specimens

Polished specimens

Jasper

HARDNESS: 7 **STREAK:** White

Occurrence

ENVIRONMENT: Lakeshore, mine dumps, riverbeds, road cuts

WHAT TO LOOK FOR: Hard, reddish stones with a waxy feel and appearance; sometimes layered with a gray metallic mineral

SIZE: Jasper occurs massively and can be found in virtually any size, though typically palm-sized or smaller

COLOR: Brown to red, white to yellow, green

OCCURRENCE: Very common

NOTES: Jasper is a very common form of microcrystalline quartz (quartz crystals too small to be seen) found easily on the lakeshore. It is closely related to chert, therefore forming as opaque, compact masses having closely spaced, granular microcrystals. As a quartz-based mineral, it exhibits the usual hallmarks of quartz, including a waxy surface feel and luster, considerable hardness, and conchoidal fracture (when struck, circular cracks appear). Its colors, which generally range from brown to red or yellow are derived from iron and distinguish it from white or gray chert. Its opacity separates it from chalcedony, a translucent microcrystalline quartz mineral.

In Michigan, especially in the Ishpeming area, "jaspilite" can be found and is widely collected and sold. Jaspilite is actually a rock called "banded iron formation" and was originally formed at the bottom of earth's ancient oceans, long before life existed on land. It consists of bright red jasper layered with gray hematite and polishes to make attractive specimens. Jasper Knob, a hill in the city of Ishpeming, is made entirely of banded iron formation.

WHERE TO LOOK: Jasper can be found nearly everywhere, but especially the lakeshore.

Concrete Driftwood Brick

Aluminum

Tile

Beach glass

Tar Sheet metal

Slag glass

Junk

HARDNESS: N/A **STREAK:** N/A

Occurrence

ENVIRONMENT: Found in all environments

WHAT TO LOOK FOR: Man-made items that "don't look right" in the environment

SIZE: Varies greatly depending on different junk

COLOR: Varies greatly

OCCURRENCE: Very common

NOTES: No matter where you are collecting, you will always come across something that just doesn't belong. While some things are obviously garbage, there are other "specimens" that aren't so clearly dismissed. Concrete, tar, brick, glass, and various bits of metal are all things that are tough enough to survive weathering and turn up on beaches as worn pieces of what we affectionately call "junk." Concrete and tar have a habit of looking like conglomerate rock, though both of these should be easy to tell from real conglomerate just by observing them visually. Brick can appear as a porous, light rock, but the colors and hardness of brick differ from similar-looking rocks. Aluminum "blobs" are light, gray pieces of metal, most likely from a soda can being beaten by the waves. Beach glass can be confused with beach-worn quartz, but pieces are often thin and uniform in thickness, which is unlikely with quartz. In addition, beach glass is softer than quartz. Slag glass is colorful, jagged glass found near old mining operations. This was a by-product of the smelting process, which separated the rock from the metal, and is actually molten rock that cooled very rapidly.

WHERE TO LOOK: As you can imagine, junk is found everywhere, but specifically the lakeshore and in mine dumps.

Greenish-gray color

Kimberlite

HARDNESS: N/A   **STREAK:** N/A

Occurrence

ENVIRONMENT: Mine dumps, quarries

WHAT TO LOOK FOR: Rock consisting of dark, greenish minerals, often containing fragments of other rocks within it

SIZE: Kimberlite is a rock and therefore can be found in specimens ranging from pebbles to boulders

COLOR: Varies; mostly greenish-brown to brown, gray, yellow

OCCURRENCE: Uncommon

NOTES: Kimberlite is a unique rock found only in specific places around the world. It is named for Kimberley, South Africa, where it was first discovered as a source of gem-quality diamonds. Kimberlite formations, called kimberlite pipes, form in vertical carrot-shaped tubes that extend from the earth's surface to deep within the earth's crust. These strange rock formations are created when molten rock containing large amounts of gases, such as carbon dioxide, begins to boil and then explodes upward in one violent, rapid eruption. This upward flow brings with it many rare and desirable minerals, including olivine, garnets, diopside, augite and, of course, diamonds, which makes kimberlite a very important rock both economically and geologically. It has a chunky, fragmented appearance that often has a greenish color. At first glance, certain specimens may resemble conglomerate, but you should be able to find grains of individual minerals, rather than just various rocks cemented together. And as for diamonds—tiny crystalline diamonds have been found in several Michigan kimberlites, though they are quite rare.

WHERE TO LOOK: The Lake Ellen Kimberlite Formation is about 15 miles northeast of Crystal Falls, but it can be hard to find.

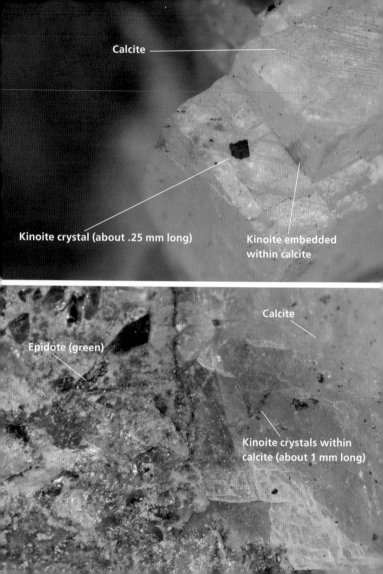

Calcite

Kinoite crystal (about .25 mm long)

Kinoite embedded within calcite

Epidote (green)

Calcite

Kinoite crystals within calcite (about 1 mm long)

Kinoite

HARDNESS: 5 **STREAK:** Bluish white

Occurrence

ENVIRONMENT: Mine dumps

WHAT TO LOOK FOR: Tiny, bright blue crystals within calcite or quartz

SIZE: Kinoite crystals are incredibly small, most often shorter than one millimeter

COLOR: Light to dark blue

OCCURRENCE: Very rare

NOTES: The discovery of kinoite in the mine dumps of the Keweenaw Peninsula copper mines was only the second known occurrence of the mineral worldwide. It is a rare mineral that is very difficult to find, despite its bright blue color. This is because Michigan's kinoites are found as tiny crystals, most often shorter than one millimeter. If that weren't enough, almost all kinoite specimens are embedded within calcite or quartz. Even though the kinoite specimens are very small, they are often quite well formed, exhibiting elongated faces ending in a blade-like point, though such details are only appreciable through a microscope. It gets its color from copper, but you won't confuse it with any other copper-based mineral.

Even average-quality kinoites can be valuable, especially given their small size, though it is not monetary gain that drives collectors to hunt for kinoite. Rather, it is the challenge of finding something so rare. Few mine dumps yield kinoite—most are in Houghton County.

WHERE TO LOOK: The Laurium Mine, in Calumet, and the Kearsarge Mine, five miles north, are two locations in Houghton County that have produced kinoite from mine dumps.

Rough Kona dolomite

Specimens courtesy of Ken Flood

Polished specimen

Kona Dolomite

HARDNESS: 4 **STREAK:** N/A

Occurrence

ENVIRONMENT: Quarries, road cuts

WHAT TO LOOK FOR: Pink rock containing fragments and darker-colored veins

SIZE: Kona dolomite forms massively and therefore can be found in any size, though most specimens are softball-sized and smaller

COLOR: Pink, brown, white to cream-colored

OCCURRENCE: Uncommon

NOTES: Kona dolomite is a variety of marble named for the Kona Hills, near Marquette, where it is found and collected. Marble is formed when limestone is metamorphosed, or changed due to heat and pressure. In this case, it was a dolomite-rich limestone, sometimes called "dolostone," that was changed to marble. The result is a soft, pink or brown rock that is a well-known Michigan collectible and has been used as a decorative stone for years. Since limestones often contain fossil material, it's no surprise that marbles can as well, including Kona dolomite. Wavy, rounded structures resembling mushrooms can sometimes be found within the stone and are actually fossils called stromatolites. Stromatolites are an ancient bacteria that still live in oceans today.

Kona dolomite isn't particularly collectible or valuable unless it has good color and lively patterns. Even then, most specimens aren't worth much unless polished or they contain well-preserved stromatolite fossils.

WHERE TO LOOK: The Kona Hills are about 8 miles southwest of Marquette and the quarries there are great places to look.

Laumontite coating

Laumontite amygdules in basalt

Calcite crystals (white)

Laumontite crystals

Laumontite

HARDNESS: 3.5–4   **STREAK:** White

Occurrence

ENVIRONMENT: Lakeshore, mine dumps, road cuts

WHAT TO LOOK FOR: Crumbly, salmon-colored fibrous crystals in or on basalt

SIZE: Individual laumontite crystals tend to be less than an inch long, while crystal aggregates can be palm-sized or larger

COLOR: Salmon to pink, orange, white to gray

OCCURRENCE: Common

NOTES: Laumontite is a member of the zeolite group (page 237), which is a family of soft, chemically complex minerals that contain water. Zeolites generally form in vesicles (gas bubbles) within basalt, and that's where you'll find all of Michigan's zeolites. Many zeolites look very much alike and form as light-colored, fibrous crystal aggregates, which can make telling them apart very difficult. Thankfully, laumontite has a few key characteristics that help identify it. First of all, laumontite's color is white when very pure, but the majority of the time you'll find it as pale orange, pink, or salmon-colored crystals. Secondly, while its hardness is 3.5, most specimens you'll find will seem much softer. This is because when laumontite is exposed to the atmosphere, the extra water in its composition evaporates, making the mineral very soft and crumbly. In fact, you can very easily break or crush most samples of laumontite with your hands. And lastly, it occurs very commonly with, or within, calcite, which can also help identify it. Some laumontite that isn't weathered may resemble a similarly colored feldspar, but even at its hardest, laumontite is still softer than all feldspars.

WHERE TO LOOK: Laumontite can be found along the shoreline of the Keweenaw Peninsula.

Limestone beach pebbles

Coral fossils

Limestone

HARDNESS: 3–4 **STREAK:** N/A

Occurrence

ENVIRONMENT: Lakeshore, riverbeds, quarries, road cuts

WHAT TO LOOK FOR: Soft, light-colored rock that often contains small fossils

SIZE: Limestone can occur in enormous beds and cliffs, but beach-worn specimens are basketball-sized or smaller

COLOR: White to gray, yellow to brown

OCCURRENCE: Very common

NOTES: Limestone is a very common sedimentary rock primarily found in the eastern side of the Upper Peninsula and in nearly all of the Lower Peninsula. This soft, light-colored rock contains over fifty percent calcite and small amounts of dolomite, clay minerals, and sometimes quartz. Its colors generally range from white to gray when fairly pure and yellow to brown when some iron is present. Much like chert, darker-colored limestone contains more organic fossil material within it. In fact, fossil coral is easily found within limestone on many of Michigan's beaches, both on Lake Superior and Lake Michigan. These fossils appear as segmented columns or fibrous circular shapes within the stone and are much easier to see when the specimen is wet.

Because limestone contains so much calcite, it is quite easy to identify. If color and hardness aren't helping you, an acid test is the best option. Place a small drop of vinegar on the surface of the stone and you should see effervescence (fizzing) as the acid begins to dissolve the rock.

WHERE TO LOOK: Enormous beds of limestone are found on the eastern end of the Upper Peninsula and in nearly all of the Lower Peninsula.

Granular magnetite

Magnetite (black) banded with quartz (white)

Specimen courtesy of A.E. Seaman Mineral Museum

Magnetite

HARDNESS: 5.5–6.5 **STREAK:** Black

ENVIRONMENT: Mine dumps

Occurrence

WHAT TO LOOK FOR: Metallic gray mineral which a magnet will stick to

SIZE: Magnetite occurs massively and can be found in any size, though it typically occurs in basketball-sized pieces and smaller

COLOR: Iron-black

OCCURRENCE: Common

NOTES: Magnetite is an ore iron that is found in several different forms, all of which are easy to identify due to the trait that gives it its name: magnetism. Magnetite looks similar to hematite and goethite, but magnetite will always attract a magnet, whereas goethite and hematite won't. Ilmenite and pyrrhotite are two other magnetic Michigan minerals, but ilmenite only exhibits weak magnetism (it only bonds loosely with a magnet) and has a browner streak while pyrrhotite isn't black in color. Magnetite commonly forms by itself as masses, sometimes with a granular, chunky texture, but it also forms as grains within quartz or chert as a low-grade iron ore called taconite. Though rare, magnetite has been found in Michigan mines in its true form—octahedral (eight-sided) crystals. And it is also a very common constituent of dark rocks, namely basalt and gabbro. Like ilmenite, it frequently weathers out of these rocks and contributes to the dark color of black sand beaches. Finally, a unique variety that contains a natural magnetic charge is called lodestone and actually acts as a magnet.

WHERE TO LOOK: The Marquette Iron Range has produced the most magnetite as well as some of the best specimens.

Massive malachite

Inset specimen courtesy of Alex Fagotti

Malachite with cuprite

Malachite crystals (dark green)
with chrysocolla (light green)
and calcite (white)

Malachite

HARDNESS: 3.5–4 **STREAK:** Light green

Occurrence

ENVIRONMENT: Mine dumps

WHAT TO LOOK FOR: Bright green coatings and crystals on copper or in copper ores

SIZE: Massive malachite can be palm-sized, though generally is smaller, and malachite crystals are normally very small, measuring just a few millimeters

COLOR: Dark green to light green

OCCURRENCE: Uncommon

NOTES: When most US mineral collectors think of malachite, they picture Arizona's beautiful banded specimens. But as a copper-based mineral, Michigan's huge native copper deposits have also produced their share of the green mineral. Crystals are rare, but they can be found in mine dumps as very small, round, ball-like groupings of tiny needles. Malachite is most often found as a dark green layer or coating atop copper or other copper-related minerals, such as chrysocolla or cuprite. These coatings are often soft and dusty, and can easily be destroyed simply by overhandling. Massive malachite is a better target for collecting, as larger pieces or thicker coatings will hold up better and often provide for more richly colored specimens. Copper may contribute to many minerals' green color, but the only other Michigan mineral you're likely to confuse malachite with is chrysocolla. Malachite is generally a darker green color, whereas chrysocolla is more of a bluish-green. But if you can't tell visually, chrysocolla is softer and is far more common.

WHERE TO LOOK: The presence of malachite signaled that copper was near, and miners took note. Look for malachite in the Keweenaw Peninsula's copper mine dumps.

Manganite crystals

Radiating fibrous crystals

Specimen courtesy of A.E. Seaman Mineral Museum

Manganite crystals on basalt

Manganite

HARDNESS: 4   **STREAK:** Reddish-brown to black

Occurrence

ENVIRONMENT: Mine dumps

WHAT TO LOOK FOR: Silvery-black sprays of fibrous, needle-like crystals

SIZE: Individual crystals can be several inches long, while crystal aggregates can be palm-sized and sometimes larger

COLOR: Gray to iron-black

OCCURRENCE: Uncommon

NOTES: As its name suggests, manganite is a manganese-based mineral, often found as well-crystallized specimens in iron deposits. Michigan has produced many world-class specimens of the mineral. The finest are stubby, blocky prisms, but it more frequently forms as dark, metallic aggregates of radiating needle-like crystals, forming "sprays." It is easy to confuse it with pyrolusite, another manganese mineral that forms nearly identical crystal sprays, but pyrolusite is harder and has a bluish streak, whereas manganite's streak is more brown in color. As mentioned above, it is very often found within iron deposits as crystals or veins, occurring alongside minerals like hematite and goethite. Therefore, look for manganite in iron mine dumps, as well as in copper mine dumps where manganite has been reported as very small crystals. Manganite is more common than one might think, but it can be very difficult to identify small specimens or those incorporated within other rocks, ores, and minerals.

WHERE TO LOOK: The Gogebic Iron Range, near Ironwood, and the Marquette Iron Range, near Ishpeming, have produced fantastic specimens.

Rough marble

Marble

HARDNESS: ≈3 **STREAK:** N/A

Occurrence

ENVIRONMENT: Quarries, road cuts

WHAT TO LOOK FOR: White, soft, coarse-grained rock that looks like calcite

SIZE: Marble forms massively and can be found in any size, from pebbles to boulders

COLOR: White to gray; black, green, red, brown, or yellow due to impurities

OCCURRENCE: Common

NOTES: Marble has been the preferred medium for sculpture for thousands of years because of its low hardness, ease of shaping, and its semi-translucent quality that can resemble human skin, a trait Michelangelo exploited in his statue of David. Marble is formed when limestone, a rock consisting of over fifty percent calcite, is metamorphosed, or subjected to great heat and pressure. This compacts the minerals within the rock, giving many marbles a coarse-grained appearance. Exceptionally pure marbles are bright white in color, but many are affected by impurities. For example, marble containing a lot of fossil material is often gray or black. Green marble is the result of inclusions of micas or serpentines, and yellow marble is created by iron oxide stains. The eastern end of the Upper Peninsula has the most limestone, so naturally it has the most marble. You're unlikely to confuse marble with any other rock, as its hardness and appearance are fairly distinctive. Kona dolomite, a collectible Michigan rock, is a form of pink, dolomite-rich marble.

WHERE TO LOOK: The quarries near Felch in Dickinson County produce large amounts of marble.

Marcasite crystals

Decaying marcasite (gray)

Marcasite

Marcasite

HARDNESS: 6–6.5 **STREAK:** Dark gray to black

Occurrence

ENVIRONMENT: Mine dumps, quarries

WHAT TO LOOK FOR: Light-colored metallic mineral often growing as tabular (plate-like) crystals or angular masses

SIZE: Individual crystals are generally smaller than your thumb-nail; masses of marcasite can be palm-sized or larger

COLOR: Light brass-yellow, white to gray

OCCURRENCE: Common

NOTES: Marcasite is an iron mineral with the same chemical composition as pyrite, but the conditions in which it forms are more acidic. This acidity makes marcasite unstable, thus pyrite is the "preferred" form taken on by this chemical composition. Because of this fact, pyrite is more common, but marcasite is by no means rare. Marcasite is an unfortunate example of a mineral that "self-destructs." Over a few years' time, it will oxidize and decompose as the sulfur in its composition is given off into the atmosphere and combines with moisture in the air to create sulfuric acid. This acid hastens marcasite's destruction, turning it to a grayish-white, crumbly material that no longer resembles the crystals you collected. Storing specimens in air-tight containers may slow this process, but there is no known "cure."

Marcasite is found in the Upper Peninsula's many iron-rich areas, such as in the mine dumps near Marquette and Ironwood. In the Lower Peninsula, it is found in shale and limestone deposits.

WHERE TO LOOK: The iron mines near Champion, 13 miles west of Ishpeming on US-41, have produced many fine crystals.

Muscovite in pegmatite

Biotite (black) in granite

Muscovite

Muscovite "books"

Quartz

Feldspar

Specimen courtesy of Michael P. Basal

Mica group

HARDNESS: 2.5–3 **STREAK:** Colorless

Occurrence

ENVIRONMENT: Road cuts, quarries, mine dumps

WHAT TO LOOK FOR: Shiny, almost metallic, dark-colored mineral that forms in thin sheets

SIZE: Most micas are thumbnail-sized or smaller, but in certain rock formations they can be palm-sized and larger

COLOR: Colorless, brown, gray to black, rarely yellow

OCCURRENCE: Common in the Upper Peninsula; uncommon in the Lower Peninsula

NOTES: The mica group is made up of many minerals that are easily identified as a mica, but they are difficult to tell apart from each other. This is because most micas are brown or black in color and have a high luster, sometimes they are so shiny that they appear metallic. They also all occur as thin, flaky crystals that form in "books," or stacks. Individual sheet-like crystal "pages" can be peeled off and separated from the rest of the book. In fact, individual crystals of mica are extremely flexible and can be bent quite a bit before they finally break. Individual crystals are very often transparent and have been used as a glass substitute for centuries.

In Michigan, the primary varieties of mica found are phlogopite, muscovite and biotite. Biotite is normally only found as black, glassy patches in granite. Phlogopite is common in kimberlite and in dolomitic marbles, such as Kona Dolomite. Muscovite is arguably the most common and most collectible mica; it is frequently collected in the pegmatite areas near Marquette as large, flat crystals.

WHERE TO LOOK: Micas are found mostly in the western half of the Upper Peninsula, especially near Republic.

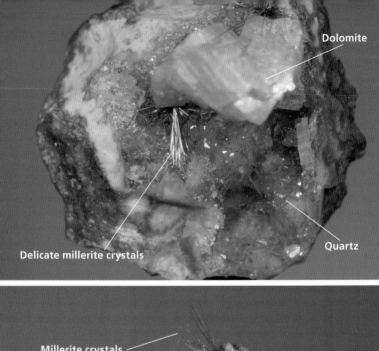

Dolomite

Delicate millerite crystals

Quartz

Millerite crystals

Dolomite

Quartz

Millerite

HARDNESS: 3–3.5 **STREAK:** Dark green

Occurrence

ENVIRONMENT: Quarries, mine dumps

WHAT TO LOOK FOR: Small, needle-like metallic crystals within geodes

SIZE: Most millerite crystals are fractions of a millimeter wide and no longer than one centimeter

COLOR: Brass-yellow

OCCURRENCE: Rare

NOTES: Millerite is a rare find in Michigan, with only a few known areas where it can be found. The tiny, needle-like crystals are composed primarily of the element nickel as well as sulfur, which gives them their brassy-yellow metallic color. Millerite is very easy to identify and generally occurs as small "sprays," or radial groupings, of the tiny crystals. If you were to look at the cross section of a millerite crystal under a very strong microscope, you would see that the needles are hexagonal, or six-sided.

Millerite has been found near the city of Ishpeming within quartz, in the same environment as the gold veins also found there. However, the finest Michigan millerite specimens come from quarries in Huron County where they are found within geodes. Geodes are small, round bodies of rock that are hollow, often containing quartz, dolomite and calcite. Millerite is still rare within them, but it is often the great "prize" for people who diligently hunt for geodes.

WHERE TO LOOK: Fantastic specimens are found within geodes or pockets of quartz from the area near Bay Port in Huron County.

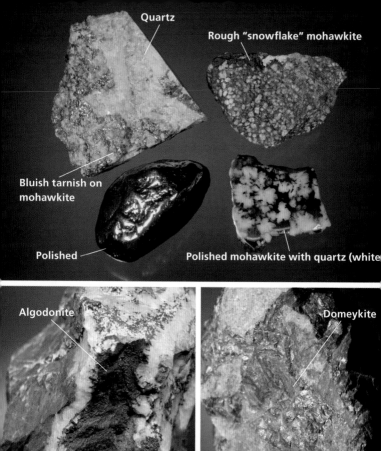

Quartz

Rough "snowflake" mohawkite

Bluish tarnish on mohawkite

Polished

Polished mohawkite with quartz (white

Algodonite

Domeykite

Specimens courtesy of   A.E. Seaman Mineral Museum

HARDNESS: Varies **STREAK:** Varies

Occurrence

ENVIRONMENT: Mine dumps

WHAT TO LOOK FOR: Yellowish-gray metallic rock often containing lots of quartz

SIZE: Mohawkite occurs in a wide range of sizes, from pebbles to boulders

COLOR: Brassy-yellow to metallic gray, sometimes with a blue iridescent surface tarnish or yellow staining

OCCURRENCE: Rare

NOTES: Mohawkite is a rare rock that actually consists of mixtures of arsenic and copper combined with quartz. Named for the Mohawk area, where it was originally found, mohawkite's two prominent ingredients are the arsenic-rich copper minerals algodonite and domeykite, which both look like brownish-yellow, brassy metals when separate. Mohawkite is therefore a similar brass color, though it sometimes develops a thin surface tarnish of blues and greens or a deep yellow stain. Most mohawkite occurs with or within quartz, and a variety called "snowflake mohawkite" contains small pockets or flecks of quartz within it. Its color can resemble that of pyrrhotite, but pyrrhotite is magnetic whereas mohawkite is not. Although mohawkite contains arsenic by way of its main ingredients, algodonite and domeykite, it is safe to handle, though you should still wash your hands afterwards and be careful never to breathe in dust created when working with it or polishing it.

WHERE TO LOOK: Mohawk is about 6 miles northeast of Calumet, on the Keweenaw Peninsula.

Quartz

Molybdenite crystal

Specimen courtesy of
Michael P. Basal

Molybdenite in schist

Molybdenite (metallic gray)

Molybdenite

HARDNESS: 1–1.5 **STREAK:** Grayish-green to black

Occurrence

ENVIRONMENT: Mine dumps

WHAT TO LOOK FOR: Small, very soft, very reflective, layered crystals or masses within quartz or rock

SIZE: Individual crystals are generally very small, being no larger than a pea

COLOR: Bluish-gray

OCCURRENCE: Rare

NOTES: Molybdenite is the primary ore of the element molybdenum, which is used to create strong alloys of steel and chrome. Molybdenite forms as small, hexagonal (six-sided) crystals that are very thin. These sheet-like crystals form in stacks, like mica, to create layered crystal groupings. Well-formed crystals are rare, however, and molybdenite is more commonly found within quartz or schist as small, irregular flakes. But whether or not crystals are present, molybdenite is very easy to identify. Its highly reflective, metallic, bluish-gray color is difficult to confuse with anything else, and its extreme lack of hardness removes any doubt. Molybdenite's streak color is very distinctive as it is grayish-green on a streak plate and dark gray to black on paper. Micas are perhaps the only minerals that could be confused with molybdenite given the similarities in crystal structure along with both minerals' great flexibility, though they are easy to distinguish. Molybdenite occurs in Michigan with quartz, chlorite, tetrahedrite, and rarely gold.

WHERE TO LOOK: The Michigan Gold District, north of Ishpeming, has scattered abandoned gold mines that have produced many specimens of molybdenite in quartz.

Gem-grade forsterite (also known as peridot)

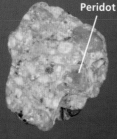

Peridot in kimberlite

Peridot removed from kimberlite

Specimens courtesy of Michael P. Basal

Olivine (green) in gabbro

Olivine group

HARDNESS: 6.5–7 **STREAK:** Colorless

Occurrence

ENVIRONMENT: Mine dumps, lakeshore

WHAT TO LOOK FOR: Hard, green crystals or pockets, most commonly within rocks, namely gabbro

SIZE: Fine-quality olivine is generally no larger than a pea, and olivine of poor quality embedded within rock is generally the same size

COLOR: Yellow-green to deep green, brown

OCCURRENCE: Very common within rocks; gem-grade grains are rare

NOTES: You'll often see minerals labelled simply "olivine," but that name actually refers to a series of hard, closely related minerals. All olivine minerals are found in shades of green or brownish-green. Forsterite is the most common variety in Michigan and it is primarily found as grains within rocks, such as gabbro, giving that rock its greenish color. More collectible specimens can be found in kimberlite as small green nuggets or masses embedded in the rock. These translucent, gem-quality forsterite specimens are called peridot and are very collectible. Peridot is rare in Michigan, but large, high-quality pieces can be cut and used in jewelry. Epidote's similar hardness and color could be confused with olivine, but epidote is often found as small, tabular (flat, plate-like) crystals within vesicles (gas bubbles) or as coatings on basalt, and olivine doesn't form in this way. When olivine weathers, it alters to (turns into) serpentine, and most olivine not found in kimberlite is at least partially serpentine.

WHERE TO LOOK: Peridot in kimberlite is rare; the Lake Ellen Kimberlite Formation is 15 miles northeast of Crystal Falls.

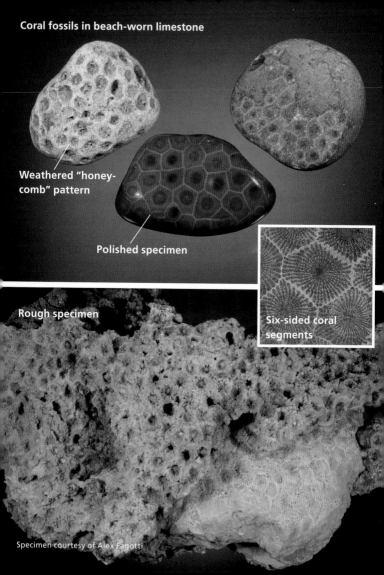

Coral fossils in beach-worn limestone

Weathered "honey-comb" pattern

Polished specimen

Rough specimen

Six-sided coral segments

Specimen courtesy of Alex Fagotti

Petoskey Stone

HARDNESS: N/A **STREAK:** N/A

Occurrence

ENVIRONMENT: Lakeshore

WHAT TO LOOK FOR: Limestone with six-sided coral fossils within

SIZE: Petoskey stones tend to be softball-sized or smaller

COLOR: White to gray, tan to brown

OCCURRENCE: Uncommon

NOTES: When it comes to famous Michigan rocks and minerals, the Upper Peninsula has copper, and the Lower Peninsula has Petoskey stone. Designated as the Michigan State Stone in 1965, Petoskey stone is a variety of limestone containing fossil coral. There are many different kinds of fossil coral found in Michigan, but Petoskey stone is one very particular species: *Hexagonaria percarinata*. This extinct coral grew at the bottom of an ancient ocean and its coral segments exhibit a six-walled structure. By very strict definitions, if your specimen does not have six-sided corals within it, it is not true Petoskey stone. This simple trait will help distinguish it from the many corals, such as horn coral, that are found in both peninsulas of the state.

While the best place to look is on the beaches, dry Petoskey stone pebbles look like simple limestone. Wetting or polishing is the only way for the stone to show off the fossils within. Some specimens are weathered to reveal a peculiar honey-comb-like pattern that can set it apart from regular limestone.

WHERE TO LOOK: Rounded beach-worn specimens are famously found on the beaches near Petoskey and Charlevoix in the northwest corner of the Lower Peninsula. Lesser-known beaches along Lake Huron in the northeastern portion of the Lower Peninsula also produce specimens, particularly near Alpena.

Porphyritic basalt

Feldspar crystals in basalt

Polished porphyritic granite

Large feldspar crystals in granite

Porphyry

HARDNESS: Varies **STREAK:** N/A

Occurrence

ENVIRONMENT: Lakeshore, road cuts, mine dumps

WHAT TO LOOK FOR: Finer-grained rock containing large, jagged crystals of feldspar

SIZE: The rocks themselves can be any size, but the porphyritic crystals are generally no more than a few inches long

COLOR: Varies greatly; porphyritic crystals are generally white, pink, or orange

OCCURRENCE: Common

NOTES: Porphyry is a rock that can be identified based on visual characteristics alone, which is an uncommon trait among rocks. It is most easily spotted on the lakeshore as a fine-grained rock, especially basalt, with jagged, rectangular crystals visible within it. These "extra" crystals are feldspars that formed separately from the rest of the rock. This happens when molten rock deep within the earth begins to cool very slowly and the feldspars within it start to crystallize. Then, the process is disturbed when the molten rock is quickly erupted, freezing the well-formed feldspar crystals in place while the rest of the rock quickly cools and fills in around them. These differ from amygdules, which are minerals that filled round, irregularly shaped vesicles (gas bubbles) within rocks long after the rocks cooled. But it's not just basalt that can get the porphyritic treatment. Granite can also commonly form as porphyry, though it is not as obvious due to granite's already coarse-grained texture. Look for stubby, square or hexagonal (six-sided) crystals that are larger than the other grains within the rock.

WHERE TO LOOK: Porphyry is most easily found on the lakeshore.

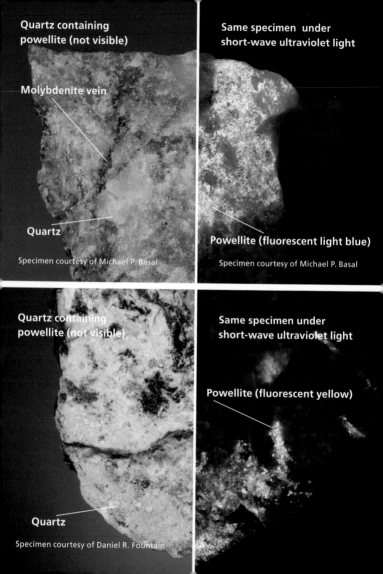

Quartz containing powellite (not visible)

Molybdenite vein

Quartz

Specimen courtesy of Michael P. Basal

Same specimen under short-wave ultraviolet light

Powellite (fluorescent light blue)

Specimen courtesy of Michael P. Basal

Quartz containing powellite (not visible)

Quartz

Specimen courtesy of Daniel R. Fountain

Same specimen under short-wave ultraviolet light

Powellite (fluorescent yellow)

Powellite

HARDNESS: 3.5–4 **STREAK:** Light yellow

Occurrence

ENVIRONMENT: Mine dumps

WHAT TO LOOK FOR: Areas of rock or quartz that fluoresce yellow- or bluish-white under short-wave ultraviolet light

SIZE: Individual powellite crystals are often too small to see without a microscope

COLOR: Green to yellow; light yellow or blue to white under short-wave ultraviolet light

OCCURRENCE: Very rare

NOTES: Powellite is one of Northern Michigan's very rare minerals and only the luckiest and most skilled collectors have even the slightest chance of finding it. The extremely rare deep-green crystal points once found in the Keweenaw Peninsula's copper mines are long gone, and your only hope of finding Michigan powellite is in the gold district near Ishpeming in Marquette County. There it forms as tiny grains spread throughout quartz along with molybdenite, though it generally cannot be seen with the naked eye. So how do you know if you've found a piece of powellite-bearing material? Hold it under a short-wave ultraviolet light and it will fluoresce a yellowish-white color if the powellite is pure. Some powellite, however, contains impurities of the element tungsten which will cause it to fluoresce a bluish-white instead. Some serious collectors will hunt for powellite at night with a short-wave fluorescent flashlight, looking for the tell-tale glow. Because of its rarity and fluorescence, you won't easily confuse powellite with other minerals.

WHERE TO LOOK: The Michigan Gold District runs from west to east, north of Ishpeming. Good collecting sites are rare.

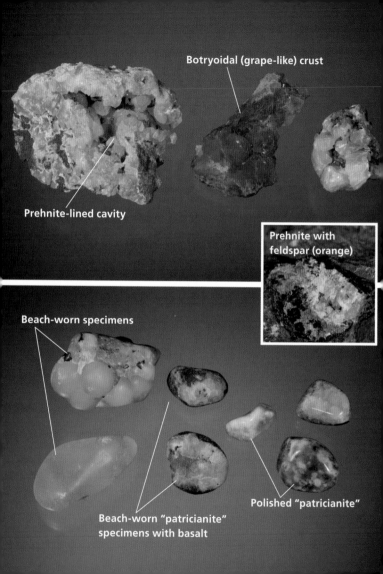

Botryoidal (grape-like) crust

Prehnite-lined cavity

Prehnite with feldspar (orange)

Beach-worn specimens

Beach-worn "patricianite" specimens with basalt

Polished "patricianite"

Prehnite

HARDNESS: 6–6.5 **STREAK:** White

Occurrence

ENVIRONMENT: Lakeshore, mine dumps, road cuts

WHAT TO LOOK FOR: Pale green crystals forming rounded crusts within pockets in basalt, or as translucent beach pebbles

SIZE: Prehnite crystal aggregates are commonly palm-sized, sometimes larger, while beach-worn specimens are often thumbnail-sized or smaller

COLOR: Light green, white to gray, rarely pink

OCCURRENCE: Common

NOTES: Prehnite is a common Michigan collectible easily found in a number of forms by amateurs and skilled collectors alike. It is virtually always found as an easy-to-spot pale apple-green or gray-green color. Individual crystals are rare; instead, it is mostly found as rounded, botryoidal (grape-like) crusts within cavities in rocks. Also common are prehnite-filled vesicles (gas bubbles) within basalt, occurring with calcite, chlorite, feldspar and copper. These hard nodules (round mineral clusters) often display a fibrous cross section that can resemble a zeolite, though prehnite is much harder than any zeolite found in Michigan.

Prehnite is frequently found on the beaches of the Keweenaw Peninsula as pale green and pink translucent pebbles that are sometimes called "patricianite." Collectors often mistake patricianite for thomsonite thanks to its pink coloration (caused by copper inclusions) and its fibrous appearance. Perhaps this is "wishful thinking" on the part of collectors, since thomsonite is a rare and valuable mineral, but, in truth, pink prehnite is actually more rare than thomsonite.

WHERE TO LOOK: Look in the Keweenaw Peninsula's mine dumps and on the lakeshore.

Pudding Stone

HARDNESS: ≈7 **STREAK:** Varies

ENVIRONMENT: Lakeshore, riverbeds

Occurrence

WHAT TO LOOK FOR: White rock containing jagged fragments of bright red jasper

SIZE: Pudding stone occurs massively and can be found in any size, from pebbles to boulders

COLOR: White to gray, with spots of red and brown

OCCURRENCE: Uncommon

NOTES: Pudding stone, like Michigan's granite, is another case of "foreign" rocks making their way south via the glaciers of past ice ages. Pudding stone formed in Canada and is a type of conglomerate rock that has undergone metamorphosis; it has been greatly compacted and hardened due to pressure. It originally consisted primarily of gravel-sized pieces of jasper, chert and quartz, though after such intense hardening and cementing together, the quartz has actually become a variety of quartzite. The jasper and chert pieces are randomly sized and shaped and are generally quite limited in relation to the amount of quartzite.

This metamorphosed conglomerate derives its curious name from its appearance. Long ago, its discoverers thought that the bright red jasper fragments resembled fruit and the dark chert looked like nuts, with the white quartzite being the "pudding" they floated in. It is this characteristic appearance that should be enough to identify it.

WHERE TO LOOK: Pudding stone is often found on the eastern end of the Upper Peninsula, particularly where the lakeshore nears the Canadian border, but they can also occasionally be found on Lake Huron's shores in the Lower Peninsula.

Polished specimen in basalt

Beach-worn specimen in basa

Beach-worn specimens

Polished chlorastrolite

Pumpellyite crystals

Quartz

Pumpellyite group

HARDNESS: 5.5–6 **STREAK:** White

Occurrence

ENVIRONMENT: Mine dumps, lakeshore

WHAT TO LOOK FOR: Olive green needle-like crystals or nodules showing a "turtle-back" pattern

SIZE: Most pumpellyite specimens are pea-sized or smaller

COLOR: Olive-green to gray-green, bluish-green

OCCURRENCE: Rare

NOTES: The pumpellyite group may have been originally discovered in Europe, but it was first described in detail from the Keweenaw Peninsula in 1920 during the heyday of Michigan's copper mining industry. Twenty years later, a variety called chlorastrolite was identified and soon after became the Michigan state gemstone. Chlorastrolite is by far the best known and most collectible pumpellyite mineral. However, it is more commonly known as "Isle Royale greenstone," due to its famous occurrence on Isle Royale, or simply as "greenstone." Chlorastrolite is found as nodules (round mineral clusters) within basalt or as beach-worn pebbles on the shore that exhibit a spiderweb-like pattern that some say resembles a "turtle's back." These rare, valuable gems are often cut and polished for use in jewelry. Other members of the pumpellyite group can be found as small, olive-green groupings of needle-like crystals within pockets in quartz and feldspars, or within basalt vesicles (gas bubbles). You could confuse it with epidote, but epidote is harder and more yellow in color.

WHERE TO LOOK: Isle Royale is considered the best source of chlorastrolite, though the island is a National Park and it is illegal to collect there. Copper mine dumps in the Keweenaw also produce the rare mineral.

Chalcopyrite

Pyrite cubes

Pyritohedron on calcite

Calcite crystals

Pyrite coating

Pyrite

HARDNESS: 6–6.5 **STREAK:** Greenish-black

Occurrence

ENVIRONMENT: Riverbeds, quarries, mine dumps, road cuts

WHAT TO LOOK FOR: Yellow metallic mineral often found as cubes

SIZE: Pyrite crystals are generally thumbnail-sized and smaller, though they can be found in much larger sizes

COLOR: Brass-yellow to brownish-yellow

OCCURRENCE: Common

NOTES: Pyrite is a very common mineral worldwide and is popular among collectors. It has the same chemical composition as marcasite—iron and sulfur—but is yellower in color and it most often occurs as cubic crystals, which is a form marcasite doesn't take. For centuries, pyrite's metallic yellow color has tricked people into thinking they had found gold, hence pyrite's nickname, "fool's gold." But the similarities end there. Pyrite is much harder, more brittle, and exceedingly more common than gold, not to mention that their streak colors are very different. As mentioned above, pyrite is often found as cubic crystals, which are widely collected. It also forms as pyritohedrons, a crystal shape named for pyrite that consists of twelve pentagonal (five-sided) faces, somewhat like a soccer-ball. If crystals are not present, pyrite can easily be confused with a few other minerals, primarily chalcopyrite and marcasite. The differences between pyrite and marcasite are briefly mentioned above, but pyrite's greenish streak and marcasite's gray streak are normally enough to tell them apart. Chalcopyrite is similar in color, but is softer and tarnishes to shades of blue.

WHERE TO LOOK: Pyrite is so abundant that any iron mine dump in Marquette, Dickinson, and Iron Counties is a great place to look.

181

Calcite (white)

Radiating, fibrous pyrolusite crystals

"Spray" of pyrolusite

Slender, shiny crystals

Pyrolusite

HARDNESS: 6–6.5 **STREAK:** Black to bluish-black

ENVIRONMENT: Mine dumps

Occurrence

WHAT TO LOOK FOR: "Sprays" (radial aggregates) of fine, needle-like silvery-black crystals

SIZE: Individual crystals are small and no more than an inch or two long, but crystal aggregates can be several inches in size

COLOR: Black to steel-gray, silvery-black, rarely bluish

OCCURRENCE: Uncommon

NOTES: Like manganite, pyrolusite is a manganese mineral found primarily in Michigan's iron mines. Unlike many minerals which are only rarely found as fine crystals, Michigan's pyrolusite is often discovered in attractive groupings of needle-like crystals arranged into "sprays," or radiating formations. These slender, metallic crystals can greatly resemble those of manganite, but manganite is softer and has a browner streak color. And as with other manganese minerals, pyrolusite will often leave black dust on your hands after you handle it. You can find pyrolusite alongside hematite and barite in mine dumps in the Marquette and Gogebic iron ranges, near Ishpeming and Ironwood, respectively. The name "pyrolusite" is often a catch-all term to describe any fibrous, metallic, manganese-bearing mineral. Many manganese minerals can resemble pyrolusite, and unless distinguishing characteristics are present, they cannot be identified without laboratory analysis.

WHERE TO LOOK: Iron mines near L'Anse in Baraga County have produced fantastic specimens, as have mines near Alberta.

Pyrrhotite crystal

Specimen courtesy of Shawn M. Carlson
Inset specimen courtesy of Shawn M. Carlson

Pyrrhotite crystal

Mining core-sample rich with pyrrhotite (brass-colored metal)

Specimen courtesy of Kennecott Eagle Minerals Company

Pyrrhotite

HARDNESS: 3.5–4.5 **STREAK:** Dark gray

Occurrence

ENVIRONMENT: Mine dumps

WHAT TO LOOK FOR: Small brownish-bronze metallic crystals that are magnetic

SIZE: Pyrrhotite crystals tend to be small, less than a quarter of an inch in length, while massive pyrrhotite can be any size

COLOR: Yellow-bronze to brown-bronze, dark brown when tarnished

OCCURRENCE: Uncommon

NOTES: Pyrrhotite is a unique iron mineral with nearly the same chemical composition as pyrite. The difference is that pyrrhotite's structure is "defective" and is missing some iron ions. What this means for the average collector is that pyrrhotite is nearly always magnetic, a trait which, combined with pyrrhotite's color and streak, is very distinctive. For example, marcasite is a very similar color but is harder, and pyrite is yellower in color and has a greenish streak, and neither is magnetic. Well-formed hexagonal (six-sided) pyrrhotite crystals are generally quite small and rare, while massive pyrrhotite is more common from Michigan's mine dumps. It is found in many areas in the Upper Peninsula, particularly in Marquette County, both in iron and gold mines. Because of its iron deficiency, however, pyrrhotite is prone to crumbling when kept in collections for long periods of time. Some Michigan mines contain huge masses of pyrrhotite, evident in core-samples taken from the rock within the mines.

WHERE TO LOOK: Pyrrhotite is widespread throughout many of Marquette County's iron mine dumps.

Quartz (white) in granite

Quartz-lined cavity

Beach-worn quartz

Quartz (white) with epidote (green)

Quartz in basalt

Chlorite (green)

Quartz crystals in basalt vesicle (gas-bubble)

Quartz

HARDNESS: 7 **STREAK:** White

Occurrence

ENVIRONMENT: All environments

WHAT TO LOOK FOR: Light-colored and very hard crystals, veins, pockets, or pebbles

SIZE: Quartz can be found in a large range of sizes; as masses larger than a basketball or crystal points smaller than a pea

COLOR: Colorless to white, brown to red, purple

OCCURRENCE: Very common

NOTES: Quartz is the single most abundant mineral on the planet, so every rock collector, amateur or professional, should know both how to identify it and the forms it takes. Quartz consists of silicon and oxygen, otherwise known as silica, which is colorless or white when pure, but it can take on a rainbow of colors depending on impurities. Well-formed quartz crystals, commonly called "rock crystals," are six-sided and are found in cavities within rock. Quartz also commonly fills vesicles (gas bubbles) and cracks within rocks, appearing as pockets or veins. Beach-worn quartz masses are found as translucent white, round pebbles. Quartz is also one of the primary ingredients in rhyolite and granite, which makes those rocks very hard and weather-resistant. The identifying features of quartz are very important to know. Aside from its very high hardness and six-sided crystal points, quartz has a glassy luster and conchoidal fracture, which means that when struck or broken, quartz cracks or breaks in a rounded, half-moon shape. All quartz-based minerals will exhibit this fracture. Finally, quartz will produce a spark when struck with a metal object.

WHERE TO LOOK: Quartz specimens are easily picked up on the Keweenaw's lakeshore and in copper mine dumps.

Hematite-rich rock

Quartz crystal coatings stained red by iron

Amethyst

Specimen courtesy of John Perona

Quartz, varieties

HARDNESS: 7 **STREAK:** White

Occurrence

ENVIRONMENT: All environments

WHAT TO LOOK FOR: Light-colored and very hard crystals, veins, pockets, or pebbles

SIZE: Quartz can be found in a large range of sizes; as masses larger than a basketball or crystal points smaller than a pea

COLOR: Colorless to gray, brown to red, purple; multi-colored

OCCURRENCE: Uncommon

NOTES: As the most common mineral, there are many forms quartz can take. Several varieties of quartz are microcrystalline, meaning that their crystal structure is too small to see. These include agate, jasper, chert and chalcedony, all of which are common throughout Michigan. Other well-crystallized varieties are named for the different colors they can take. Amethyst is a purple variety of quartz that contains impurities of iron and aluminum. It is very uncommon in Michigan, but can occassionally be found on the Keweenaw Peninsula. Gray or black quartz, colored by aluminum, is more common and appears as dark, or "dirty," quartz. Red, iron-stained quartz is frequent down in iron mines, especially around Marquette, but the mining operations mean that specimens rarely make it into collectors' hands. The Keweenaw Peninsula is one of the best places in Michigan to look for quartz, agate and jasper. The many mine dumps, for example, are loaded with quartz specimens, as is the lakeshore. And the Lower Peninsula produces large amounts of chert within limestone deposits.

WHERE TO LOOK: Mine dumps on most of the Upper Peninsula will yield quartz and its varieties, particularly red and gray quartz. Pieces of amethyst sometimes turn up on lakeshores.

Pudding Stone

Jasper fragments

Quartzite

Beach-worn specimens

Quartzite

HARDNESS: ≈7 **STREAK:** N/A

Occurrence

ENVIRONMENT: Lakeshore, quarries, riverbeds, road cuts

WHAT TO LOOK FOR: Hard, grainy, light-colored rock that shares the traits of quartz

SIZE: Quartzite occurs massively and can be found in any size, from pebbles to boulders

COLOR: White to gray, yellow to brown

OCCURRENCE: Common

NOTES: Quartzite is a metamorphic rock that forms when grains of sand rich in silica (quartz material) are heated and compacted. This causes the individual grains of sand to cement together and recrystallize as one mass of quartzite. Sandstone, which is a rock consisting entirely of sand, is generally the rock from which quartzite forms. Telling the two apart is easy, however, since sandstone is generally more colorful, gritty-feeling, and is more loosely cemented together—you can easily pull individual grains from it. Quartzite, on the other hand, is normally found in shades of white, gray, or yellow, and while it can also be grainy in appearance, it is much harder and compact than sandstone. It is more likely that you would confuse quartzite with regular quartz, but quartzite is more opaque and dense. Pudding stone (page 177) is a famous variety of conglomerate rock found on the eastern shores of the Upper Peninsula. It consists of quartzite formed from quartz pebbles that have been recrystallized and cemented into one hard mass. Within the quartzite are fragments of bright red jasper and dark chert which make the entire rock look like a holiday pudding.

WHERE TO LOOK: Look on the beaches east of Grand Marais, but specifically near the lakeshore and in mine dumps.

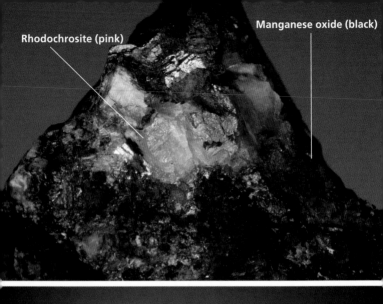

Rhodochrosite (pink)

Manganese oxide (black)

Rough rhodochrosite (pink)

Rhodochrosite

HARDNESS: 3.5–4 **STREAK:** White

Occurrence

ENVIRONMENT: Mine dumps

WHAT TO LOOK FOR: Pink, soft, blocky crystals or masses

SIZE: Rhodochrosite crystals can range from pea-sized to palm-sized, with crystal aggregates being larger

COLOR: Light to dark pink; darkens to brown on exposure to air

OCCURRENCE: Rare

NOTES: Rhodochrosite is a very collectible, very identifiable manganese mineral that, unfortunately, is quite rare in Michigan. Like its mineral cousins siderite and calcite, it forms rhombohedral crystals (crystals resembling a leaning cube) when well developed. Otherwise, it is commonly found as soft, pink masses that, when carefully struck, break into perfect rhombohedrons. This is called cleavage, or the action of a mineral breaking along the planes of its crystalline molecular structure. This type of cleavage is exactly the same as in calcite or siderite, though rhodochrosite's color differentiates it from siderite and its hardness distinguishes it from calcite. Sometimes, manganese-rich calcite will be a similar pink color, but calcite, no matter what color, is always softer.

As with most Michigan manganese minerals, rhodochrosite is found in iron mines alongside iron-based minerals, especially hematite. Its bright rose-colored crystals are famous and highly sought-after by collectors all over the world, though the finest rhodochrosites come from Colorado. Michigan's specimens, however, are prized for their rarity.

WHERE TO LOOK: Iron mines in Iron County produced some of the best specimens, but rhodochrosite is quite rare in the mine dumps.

Beach-worn rhyolite

Banded rhyolite

Inset specimen courtesy of Alex Fagotti

Vesicular rhyolite

Banded rhyolite

Rhyolite

HARDNESS: 6–6.5 **STREAK:** N/A

Occurrence

ENVIRONMENT: Lakeshore, riverbeds, road cuts

WHAT TO LOOK FOR: Fine-grained, reddish or grayish rock, often with many vesicles (gas bubbles) and colored bands

SIZE: Rhyolite can be found in any size, from pebbles to boulders

COLOR: Gray, red to brown, multi-colored bands

OCCURRENCE: Very common

NOTES: Rhyolite, like basalt, is a volcanic rock that formed when molten rock, or lava, was spilled onto the earth's surface. Exposure to the atmosphere caused it to cool very rapidly, which prevented individual minerals within it from crystallizing enough to be visible. This is in contrast to rocks like granite which cooled very, very slowly, deep within the earth, allowing the different minerals within it to crystallize, creating large, coarse, quite visible crystals. Actually, granite and rhyolite are essentially the same exact rock; they both contain the same minerals, such as quartz, light-colored feldspars, and amphiboles. The primary difference between rhyolite and granite is simply the length of time they cooled.

Rhyolite is very easily found on the Keweenaw Peninsula's beaches, primarily as reddish, fine-grained rocks. Many specimens also exhibit banding or layering, which is evidence that rhyolite was flowing while still molten. Some samples contain many vesicles (gas bubbles)—sometimes more so than other volcanic rocks. This occurs because molten rhyolite is much thicker than other types of lava (such as basalt), and it traps gases more easily.

WHERE TO LOOK: Many of the Keweenaw Peninsula's beaches appear red from afar due to their ample supply of rhyolite.

Layered sandstone

Various sandstones

Layered habit

Sandstone

HARDNESS: N/A **STREAK:** N/A

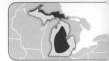

Occurrence

ENVIRONMENT: Lakeshore, riverbeds, quarries, road cuts

WHAT TO LOOK FOR: Rocks that occur in layers and have a rough, gritty feel

SIZE: Sandstone can occur in any size, from cliffs to pebbles

COLOR: White to gray, yellow to brown or tan, red, orange

OCCURRENCE: Very common in the Lower Peninsula; common in the Upper Peninsula

NOTES: Sandstone is a rock that is very easy to identify because it is exactly what it sounds like—a rock made from sand. It forms when large beds of sand, generally at the bottom of bodies of water, compress and solidify. Because it is a sedimentary rock, sandstone can be very highly layered, visible as bands of color throughout the stone, and these layers sometimes weather and split apart. Sandstone is found in shades of red, orange, brown, yellow, and cream-colors. Sand is primarily silica, or quartz, so you may think that sandstone would be a very hard rock. In reality, while the individual grains of sand may be quite hard, sandstone itself is normally very loosely held together. In fact, you can generally pull individual grains of sand off the rock.

When sandstone becomes compacted and cemented together by solutions of silica (quartz), it becomes quartzite, a much harder rock. Confusing the two rocks is unlikely, because quartzite resembles quartz much more than it resembles sandstone.

WHERE TO LOOK: The central portion of the Lower Peninsula has large beds of sandstone, but samples can also be found on the shorelines of the Upper Peninsula.

Saponite (white)

Quartz (colorless)

Epidote (green)

Basalt (brown)

Epidote (green)

Saponite "fuzz ball"
(about .5mm across)

Saponite

HARDNESS: 1.5–2 **STREAK:** White

Occurrence

ENVIRONMENT: Mine dumps, road cuts

WHAT TO LOOK FOR: Tiny white "fuzz balls" in basalt vesicles (gas bubbles)

SIZE: Saponite crystal clusters are most often very small and measure less than one millimeter across

COLOR: White to cream-colored, gray, green

OCCURRENCE: Uncommon

NOTES: Saponite is a member of the smectite group, which is a family of clay minerals. When most people think of clay, they think of the sticky, thick mud found in riverbanks, but within that wet silt are microscopic crystals of clay minerals. There are two primary groups of clays found in Michigan—the smectite group and the kaolinite group. Most cannot be identified by anything short of laboratory analysis, but saponite can sometimes be identified visually and make for interesting, albeit tiny, specimens. Most of the best specimens are found in the mine dumps of the Keweenaw Peninsula copper mines and appear as tiny white "fuzz balls" within cavities in basalt. These delicate little clusters of needle-like crystals are generally very small, measuring no more than one millimeter across. Indeed the biggest challenge of finding these crystals is simply *seeing* them. When the mineral olivine weathers and deteriorates it can turn into saponite, sometimes retaining olivine's original appearance. This is called a pseudomorph, or a mineral that looks like another mineral. A hardness test can help identify these.

WHERE TO LOOK: Copper mine dumps near Calumet in Houghton County produce the delicate white "fuzz balls."

Chlorite schist

Garnets

Schist

Chloritoid

Schist

HARDNESS: N/A **STREAK:** N/A

Occurrence

ENVIRONMENT: Road cuts, mine dumps, quarries

WHAT TO LOOK FOR: Layered, compact rock that appears to be made of many thin sheets and is often "glittery"

SIZE: Schist occurs massively and can be found in any size, from pebbles to boulders

COLOR: Gray to black, green, bluish

OCCURRENCE: Common

NOTES: When a rock is subjected to great heat and pressure, it can change completely—this is called metamorphism. What commonly happens is that the various minerals within the rock begin to order themselves into layers. The more layers a rock has and the tighter the spacing between those layers, the more metamorphism it has undergone. Schist is an example of a rock that has been completely metamorphosed and no longer resembles the original rock. It has tightly packed layers that are often made up mostly of a single mineral, especially mica. Chlorite schist is common in Michigan and is made mostly of the mineral chlorite, which makes the rock a dark green color. When schists are in the process of forming, the great heat and pressure also separate various minerals and elements within the rock, organizing and concentrating them into pockets. These concentrations can often be very hard, rare, and feature collectible minerals, such as garnets, staurolite and andalusite. Gneiss (page 113) is very similar to schist, but is much less layered and still resembles the original rock.

WHERE TO LOOK: The Lake Michigamme area, about 40 miles west of Marquette on US-41, has large amounts of schist.

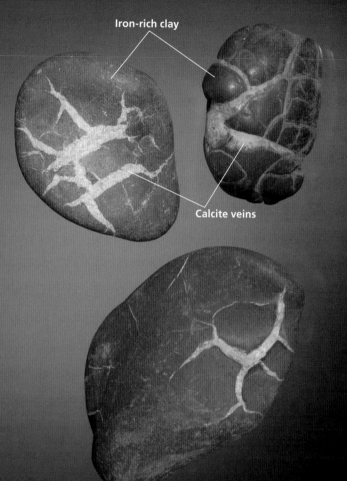

Septarian geodes

Iron-rich clay

Calcite veins

Septarian Geodes

HARDNESS: Varies **STREAK:** N/A

Occurrence

ENVIRONMENT: Quarries, road cuts

WHAT TO LOOK FOR: Rounded masses of dense, brown rock containing cracks filled with calcite

SIZE: Septarian geodes are generally no larger than a golf ball

COLOR: Varies; reddish-brown on the outside, white to yellow cracks inside

OCCURRENCE: Rare

NOTES: Geodes generally are round, hollow bodies of rock that can contain crystals within. Septarian geodes are sometimes called septarian nodules and they differ from average geodes in that they are not hollow—nor are they actually geodes. They do, however, still form as rounded masses that contain different minerals within them. The word "septarian" stems from the Latin word for "partition," an allusion to septarian geodes' cracked and segmented appearance. As mentioned above, septarian geodes are not geodes at all, but rather they are something called a concretion. Concretions are hard round concentrations of materials like clay, sand, or iron. Michigan's septarian geodes consist of a very fine-grained, iron-rich clay, often called "ironstone," that has spiderweb-like cracks filled with white or yellow calcite within. No one is certain what causes these cracks, but it is hypothesized that when the material was still forming, it quickly dried out, causing the inside of the concretion to shrink and crack. Calcite then later filled in these cracks.

WHERE TO LOOK: Michigan's septarian geodes come from near Glenn, in Allegan County.

Rough serpentine

Antigorite

Chrysotile

Serpentine

Inset specimen courtesy of John Perona

Chrysotile (fibrous) serpentine

Loose fibers

⚠ Serpentine group

HARDNESS: 3–5 **STREAK:** White

Occurrence

ENVIRONMENT: Mine dumps, quarries, road cuts

WHAT TO LOOK FOR: Green or yellow mineral with a distinct "greasy" feel, often with visible fibers

SIZE: Serpentine specimens are frequently softball-sized, but are generally smaller when collected

COLOR: Yellow to green, olive-green, dark green to black

OCCURRENCE: Uncommon

NOTES: The serpentine group is a closely related family of magnesium-bearing minerals. The primary serpentine minerals found in Northern Michigan are antigorite and chrysotile, both of which can occur in the same specimen. Serpentines are easily identified by their yellow-green or olive-green hues, their low hardness, massive habit (they form as chunks, rather than crystals), and by their distinct soapy or greasy texture. Antigorite forms in massive translucent green layers or veins while chrysotile tends to be more yellow in color and has a fibrous structure that feels silky to the touch. The Michigan Gold District near Marquette is well known for producing large amounts of serpentine. When more than one variety of serpentine occurs in the same specimen, it is technically considered a rock called serpentinite, which frequently occurs with talc. As mentioned above, chrysotile is a fibrous and silky variety of serpentine. Individual fibers are extremely flexible and can be peeled off the specimen. This is known as asbestos, and it can be cancerous if inhaled, so take great care when working with chrysotile.

WHERE TO LOOK: The old gold mines near Ishpeming in Marquette County have produced many fine specimens.

"Nonesuch" shale

Layered sheets

"Nonesuch" shale

Shale

HARDNESS: <5.5 **STREAK:** N/A

Occurrence

ENVIRONMENT: Riverbeds, quarries, road cuts

WHAT TO LOOK FOR: Fine-grained rock occurring in large sheets or layers

SIZE: Shale occurs in very large sheets or beds, sometimes miles across, but small specimens are generally palm-sized

COLOR: Tan or yellow to brown, black

OCCURRENCE: Common

NOTES: Shale is a soft sedimentary rock that forms in large, layered beds. The individual layers are generally quite thin and easily weather, coming apart in sheets. Different shales are made of different materials, but most are a result of mud, clay, or silt compacting and solidifying to form a rock. Clay minerals, as well as tiny grains of quartz and micas, are the primary ingredients of shale, but it can also contain high amounts of organic material, such as plants, in the form of fossils. The more organic material contained in the shale, the darker its color becomes. Some shale is so rich with fossils that it can be processed for coal and oil. Shale is fairly easy to identify from other sedimentary rocks because of its fine-grained texture, layered habit, and thin sheets.

In Michigan, shale is prevalent wherever sedimentary rocks are present. One of the most prominent formations is called the Nonesuch Shale Formation, found in the western corner of the Upper Peninsula. The best outcroppings are seen in the Porcupine Mountains State Park, where rivers have cut through the shale to create impressive layered cliffs.

WHERE TO LOOK: There are large shale formations all over the Lower Peninsula.

Massive siderite

Massive siderite

Siderite

HARDNESS: 3.5–4 **STREAK:** White to yellow

ENVIRONMENT: Quarries, mine dumps

WHAT TO LOOK FOR: Light brown, blocky masses occurring with other iron minerals

SIZE: Siderite is generally found in palm-sized pieces or smaller, though it can sometimes be much larger

COLOR: Light to dark brown

OCCURRENCE: Common

NOTES: Siderite is a light brown iron-bearing mineral common in iron-rich regions of Michigan. Well-formed, hexagonal (six-sided) crystals are very rare in the state, and siderite is mostly found as compact, massive pieces that, when carefully struck, break into perfect rhombohedrons (a shape resembling a leaning cube). This is called cleavage, or the action of a mineral breaking along the planes of its crystalline molecular structure. This type of cleavage is exactly the same as in calcite, to which siderite is related, but the opaque, brown colors of siderite distinguish the two minerals. Siderite could possibly be confused with limonite, but it is unlikely. Limonite does not have the glassy luster or rhombohedral cleavage of siderite.

Siderite forms in the iron ranges near Ironwood and Iron River (Upper Peninsula), where it can be found in mine dumps. In the Lower Peninsula, siderite forms in concretions (rounded, compact masses) with clay minerals in shale.

WHERE TO LOOK: From Charlevoix County down to Eaton and Branch Counties, siderite is found as rounded concretions in shale throughout the Lower Peninsula. Masses are also found in Marquette County iron mines (Upper Peninsula).

Float silver

Silver crystal

Vein silver

Silver

Quartz

Silver

HARDNESS: 2.5–3 **STREAK:** Silver-gray to white

Occurrence

ENVIRONMENT: Mine dumps, lakeshore

WHAT TO LOOK FOR: White silver metal, often with a dark gray surface tarnish

SIZE: Most silver specimens are thumbnail-sized or smaller

COLOR: Silver-white, dark gray when tarnished

OCCURRENCE: Rare

NOTES: Silver was once common in the Keweenaw Peninsula's copper mine dumps, but due to decades of collecting, it is considerably more difficult to find today. The Keweenaw Peninsula still is one of the only places in Michigan to find the mineral, however, and diligent collectors will have to spend great lengths of time digging in the piles to find even one small specimen. Though once you do find silver, it is quite easy to identify. Its low hardness and metallic white or gray color are distinctive, though most natural, uncleaned silver specimens will be coated with a dark gray or black tarnish. This black tarnish is actually a silver-based mineral called acanthite and it is the same tarnish that forms on silver jewelry. Scratching through this surface coating will reveal the true color below which will distinguish it from copper, which will also often be coated in a similarly colored tarnish. In fact, wherever there is copper, there is frequently silver, sometimes within the same specimen. Silver forms as veins in rocks or in minerals like quartz; rarely, it forms branching, tree-like crystals. It can also be found as float metal, which are rounded nuggets that have been shaped by glaciers.

WHERE TO LOOK: Copper mine dumps in the Keweenaw Peninsula are good places to look, though no locations are sure bets.

Limestone matrix

Calcite

Sphalerite

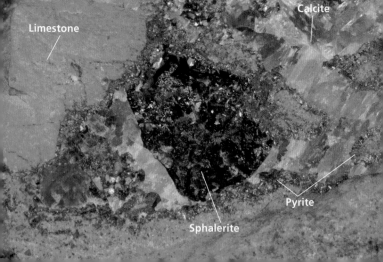

Limestone

Calcite

Sphalerite

Pyrite

Sphalerite

HARDNESS: 3.5–4 **STREAK:** Light brown

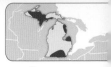
Occurrence

ENVIRONMENT: Quarries, road cuts, mine dumps

WHAT TO LOOK FOR: Dark-colored crystals, often with galena, embedded within rock

SIZE: Most sphalerite crystals are thumbnail-sized or smaller

COLOR: Brown, red, green, black

OCCURRENCE: Uncommon

NOTES: Sphalerite is the world's primary ore of zinc, though in Michigan it is quite uncommon. Well-formed crystals are pyramid-shaped or dodecahedral (twelve-sided), though these are rare in Michigan, and sphalerite is most often found as dark masses within its matrix, or host rock. Some of the best specimens are found embedded in limestone in the sedimentary-rich eastern end of the Upper Peninsula and in much of the Lower Peninsula. Specimens are generally brown, turning darker or even black when iron is present in its composition. It often occurs with galena and pyrite, though sphalerite's dark, glassy crystals are easily distinguished from those minerals, as is its hardness and streak.

A particularly fine occurrence of sphalerite is the limestone deposits in Eaton County, where it occurs with calcite and pyrite, embedded in limestone. The quarries there turn up well-formed crystals that delight Michigan mineral collectors.

WHERE TO LOOK: Limestone quarries near Bellevue in Eaton County (Lower Peninsula) produce some of the best-formed sphalerite in the state. The Michigan Gold District, near Ishpeming (Upper Peninsula) is also known for it.

Crossed crystals

Staurolite crystals
embedded in schist

Diamond-shaped
cross section

Loose staurolite crystals

Staurolite

HARDNESS: 7–7.5 **STREAK:** White to gray

Occurrence

ENVIRONMENT: Road cuts, mine dumps

WHAT TO LOOK FOR: Hard, dark crystals embedded within rock, namely schist

SIZE: Staurolite crystals tend to be thumbnail-sized and smaller, though they can be larger

COLOR: Brown, reddish-brown, gray

OCCURRENCE: Uncommon

NOTES: Like garnet and andalusite, staurolite forms in schist, where it occurs as small but very hard crystals. Staurolite is so hard, in fact, that a hardness test alone is enough to distinguish it from the vast majority of Michigan minerals. Garnet and andalusite, as mentioned above, also form in a similar rock environment and therefore are just about the only minerals hard enough to be confused with staurolite. And since their colors can all be similar, knowing staurolite's crystal form is key to distinguishing them. Staurolite forms as flat, rectangular crystals with a diamond-shaped cross section. Garnets tend to form in round, multi-sided crystals whereas andalusite is rarely found as well-formed crystals at all. In addition, staurolite's crystals often exhibit a very unique trait: two crystals frequently form through each other at perfect 90-degree angles, forming a cross or X-shape. This is called crystal twinning, and staurolite's cruciform (cross-shaped) crystals are a classic example of the process. Finely twinned specimens are prized by collectors as well as clergy.

WHERE TO LOOK: The south side of Lake Michigamme, about 40 miles west of Marquette on US-41, is known for crystals.

Sulfur fragments

Cluster of sulfur crystals

Sulfur

HARDNESS: 1.5–2.5 **STREAK:** Colorless

Occurrence

ENVIRONMENT: Quarries, road cuts

WHAT TO LOOK FOR: Bright yellow, soft, pyramid-shaped crystals

SIZE: Sulfur is often no larger than your palm, with most pieces being thumbnail-sized

COLOR: Lemon-yellow

OCCURRENCE: Rare

NOTES: Like copper and silver, sulfur is a native element found in Michigan. When an element is "native," it means that it is uncombined with other elements. For example, the mineral pyrite is the combination of one iron atom and two sulfur atoms, but native sulfur is made of nothing but sulfur atoms. Sulfur differs from native copper, silver, gold and iron because it is a nonmetallic element that is found in its native state, which is rare. It forms in vugs (cavities) within limestone and is very easy to identify because it is always a lemon-yellow color, occurs in translucent glassy crystals or masses, and is very soft. Very fine crystals are pyramid-shaped and form in clusters that can go for high prices. Sulfur occurs with calcite, celestine and gypsum, but can't be confused with any of them.

Sulfur is a very important mineral, both economically and geologically, and is common worldwide though rare in Michigan. Economically, it is used to make fertilizer and sulfuric acid. Geologically, it combines with other elements to form different minerals, several of which are in this book, including pyrite, millerite and galena, among many others.

WHERE TO LOOK: Limestone quarries in Monroe and Wayne Counties, near Detroit, have produced some of the country's finest sulfur specimens.

Soapstone

Rough talc

"Soapy" texture

Talc

HARDNESS: 1 **STREAK:** White

ENVIRONMENT: Mine dumps

Occurrence

WHAT TO LOOK FOR: Extremely soft, green mineral with a "soapy" feel that is easily scratched by your fingernail

SIZE: Talc occurs massively and can be found in a wide range of sizes, though most is commonly softball-sized and smaller

COLOR: Apple-green to mint-green, white to gray

OCCURRENCE: Uncommon

NOTES: Talc is a mineral that only forms as a result of magnesium-rich minerals, such as dolomite or serpentine, being altered by other chemicals entering their composition. Therefore, talc is found in the Upper Peninsula within the large serpentine deposits in the Michigan Gold District, near Ishpeming. Though talc's green colors can be similar to serpentine's, it couldn't be easier to tell the two minerals apart: talc is far softer. In fact, on the Mohs Hardness Scale, talc is the quintessential "1," it is always portrayed as the softest mineral on earth. What this means is that you can scratch talc with your fingernail with very little effort. This extreme softness also gives talc a soapy or greasy feel, sometimes it even feels slippery.

A variety of schist that is rich in talc, called soapstone, is also found in the same regions where talc and serpentine are found. This soft, soapy-feeling rock has been used for carvings for centuries because it is easy to shape. It is found in similar environments as talc, though tends to be whiter in color.

WHERE TO LOOK: The abandoned gold mines near Ishpeming in Marquette County have produced many fine specimens.

Massive tenorite

Specimen courtesy of A.E. Seaman Mineral Museum

Tenorite coating on copper

Specimen courtesy of John Perona

Tenorite

HARDNESS: 3.5–4 **STREAK:** Black

Occurrence

ENVIRONMENT: Mine dumps

WHAT TO LOOK FOR: Black metallic mineral, most often found coating copper

SIZE: Tenorite occurs massively and can be found in any size, though it is primarily found in palm-sized pieces or smaller

COLOR: Dark gray to black

OCCURRENCE: Common

NOTES: Just as a gray tarnish (called acanthite) forms on the surface of silver, a dark, black coating forms atop copper. That coating is called tenorite. Tenorite is a simple mixture of copper and oxygen that forms as native copper weathers and deteriorates. Tenorite is most often found as a surface coating on nearly all natural copper specimens, forming when the copper is exposed to the atmosphere. It frequently occurs with other copper-based minerals, such as cuprite, malachite, or chrysocolla. Massive pieces are rare, as they no doubt required a large concentration of copper to be exposed and weathered in order to form.

Fine examples of tenorite can have a metallic luster and resemble other black metallic minerals, such as hematite or goethite. This is unlikely, however, as tenorite is always associated with copper minerals, and simple observation will reveal the reddish-metal color of copper, the greenish tint of chrysocolla, or the glassy red crystals of cuprite. If all else fails, test for streak color and hardness.

WHERE TO LOOK: Nowhere will tenorite be easier to find than coating a specimen of Keweenaw copper in a mine dump.

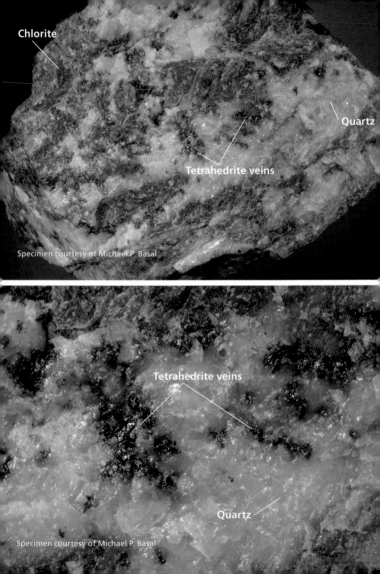

Chlorite

Quartz

Tetrahedrite veins

Specimen courtesy of Michael P. Basal

Tetrahedrite veins

Quartz

Specimen courtesy of Michael P. Basal

Tetrahedrite

HARDNESS: 3.5–4 **STREAK:** Black to brown

Occurrence

ENVIRONMENT: Mine dumps

WHAT TO LOOK FOR: Metallic black veins within quartz and schist

SIZE: Tetrahedrite specimens tend to be small and less than an inch in size

COLOR: Steel-gray to black, sometimes with blue iridescent surface tarnish

OCCURRENCE: Rare

NOTES: Tetrahedrite is a common mineral worldwide, but is rare in Michigan, and is only found in a few localities. There are several varieties, including antimony- and arsenic-rich varieties, though all are very closely related and telling them apart is very difficult. Its name is derived from its perfect tetrahedral (pyramid-shaped) crystals, though such well-formed tetrahedrite samples do not come from Michigan. Instead, it is found as black, metallic veins within quartz or chlorite schist. It will also sometimes have a bright bluish tarnish which can help differentiate it from other metallic minerals that form in the same kinds of rock, such as molybdenite. Several mines in Marquette County, including two former gold mines, have produced this material in small amounts. The famous Ropes Gold Mine, near Ishpeming, is one such mine.

Tetrahedrite's rarity and small size don't make it widely sought-after or very collectible in Michigan, but particularly fine specimens may be valuable or interesting. It may be hard to identify, but its limited occurrences help in determining it.

WHERE TO LOOK: Mines in the Michigan Gold District, north of Ishpeming in Marquette County, are good places to look.

Thomsonite in basalt

Polished specimen

Thomsonite

HARDNESS: 5–5.5 **STREAK:** White

Occurrence

ENVIRONMENT: Lakeshore, road cuts

WHAT TO LOOK FOR: Pink, fibrous crystals filling vesicles within basalt

SIZE: Most thomsonite specimens are pebbles smaller than your thumbnail

COLOR: White to pink, often with green

OCCURRENCE: Rare

NOTES: Thomsonite is a rare member of the zeolite family of minerals, which is a group of closely related, hydrous (water-bearing) minerals with very complex chemical compositions. All of Michigan's zeolites form within vesicles (gas bubbles) in basalt, and many have a fibrous appearance due to the minerals' fine, needle-like crystals that are commonly arranged into "sprays," or radial groupings. Thomsonite is no exception to this description and forms so densely within the vesicles that they often fill the entire cavity, creating a nodule (round mineral cluster). These nodules are often white or pink and contain "eyes," or circular structures. Each eye is actually the cross section of an individual thomsonite crystal cluster. These eyes continued to grow outward until they all converged to form a nodule, as mentioned above. Thomsonite may be more famous as a Minnesota mineral, but, in truth, Isle Royale produces more thomsonite than Minnesota's shores. Isle Royale, however, is a National Park and it is illegal to collect anything from the island. In the Keweenaw Peninsula, thomsonite is found as pale beach-worn pebbles, not to be confused with pink prehnite (page 175).

WHERE TO LOOK: The Keweenaw Peninsula's lakeshores are best.

225

Topaz (green-yellow) in pegmatite

Topaz

Quartz

Feldspar

Specimen courtesy of Shawn M. Carlson

Mica

Quartz

Topaz (green) in pegmatite

Specimen courtesy of Shawn M. Carlson

Topaz

HARDNESS: 8 **STREAK:** White

Occurrence

ENVIRONMENT: Quarries, road cuts

WHAT TO LOOK FOR: Very hard, greenish crystals within pegmatite (very coarse granite formation) rock

SIZE: Topaz specimens tend to be smaller than a golf ball, with most being thumbnail-sized

COLOR: Yellow to green-yellow

OCCURRENCE: Very rare

NOTES: Trace amounts of topaz have been found throughout the years in Michigan, leading many collectors to wonder if there were larger samples to be had. In recent years, this question was answered when a pegmatite formation (an outcropping of well-crystallized rock) with massive yellowish-green topaz was discovered in Marquette County. The mineral is extremely hard, which, combined with its occurrence only in pegmatite rock, is enough to identify the mineral. Fluorite occasionally forms in Michigan's pegmatites and can appear similar, but is rarely a similar green color and is much softer.

Well-formed crystals of topaz are not present and it instead appears as masses alongside quartz, micas and feldspars. Similar specimens from elsewhere in the United States would hardly be worth mentioning, but as the Black River Pegmatite Formation is such a rare, unique occurrence in Michigan, even the most average of specimens are of great interest to collectors.

WHERE TO LOOK: The Black River Pegmatite Formation in Marquette County, north of the town of Republic, is the only known occurrence.

Schorl

Quartz

Mass of schorl crystals

Specimen courtesy of A.E. Seaman Mineral Museum

Schorl (black) with mica (metallic luster)

Tourmaline group

HARDNESS: 7–7.5 **STREAK:** White

Occurrence

ENVIRONMENT: Quarries, mine dumps, road cuts

WHAT TO LOOK FOR: Hard, dark-colored, long and slender crystals embedded within other minerals or rocks

SIZE: Most tourmaline crystals are just a millimeter or two wide and less than an inch long, though they can rarely be up to an inch wide and several inches long

COLOR: Black

OCCURRENCE: Uncommon in the Upper Peninsula; rare in the Lower Peninsula

NOTES: The tourmaline group is a family of hard, brittle minerals composed of a complex combination of boron, aluminum, silicon, and a host of other elements that can turn members of this group a wide variety of colors. Michigan's most common tourmaline, however, is schorl, an iron-rich variety that appears as long, slender, black crystals. There really aren't any other Michigan minerals that you could confuse with a well-formed schorl crystal. It is always black, forming needle-like crystals with striated (grooved) faces, sometimes tapering at one end. The cross section of a schorl crystal can be described as a "bulging triangle," meaning that it resembles a triangle with outwardly rounded sides. If crystals are not present, however, rely on hardness and color.

Schorl can be found embedded in other rocks and minerals, particularly within quartz or pegmatite formations. In pegmatite occurrences, it can occur with quartz, feldspar and mica.

WHERE TO LOOK: The iron mines near Champion (Upper Peninsula) have produced fantastic schorl crystals.

Tyuyamunite (yellow) coating

Specimen courtesy of Shawn M. Carlson

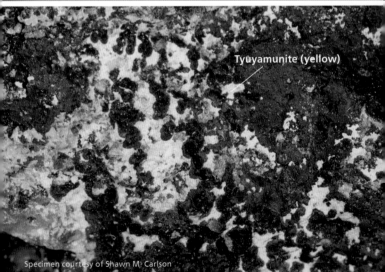

Tyuyamunite (yellow)

Specimen courtesy of Shawn M. Carlson

⚠ **Tyuyamunite**

HARDNESS: 1.5–2 **STREAK:** Light yellow

Occurrence

ENVIRONMENT: Mine dumps, road cuts

WHAT TO LOOK FOR: Bright yellow coatings of tiny, thin crystals

SIZE: Tyuyamunite coatings tend to be very small, thin patches measuring just millimeters across

COLOR: Bright yellow, canary yellow

OCCURRENCE: Very rare

NOTES: Besides having a tongue-twister of a name, tyuyamunite (pronounced tuh-YOO-ya-moon-ite) is a relatively new find in Michigan, which is exciting for Michigan mineral collectors. It is a member of the carnotite group, which is a family of bright, canary-yellow minerals that are high in uranium content. As such, tyuyamunite is radioactive, and appropriate care should be taken when collecting and storing the mineral. Luckily for cautious collectors, the mineral only occurs in small amounts as thin crusts. Visible crystals are essentially non-existent in Michigan; instead, tyuyamunite is collected as flaky, very thin, yellow coatings on the surface of other minerals or rocks. It forms from weathering uraninite, another uranium-bearing mineral that is found in Michigan only as grains within rocks like granite and slate. Tyuyamunite's rare occurrence, low hardness, and bright yellow color are always enough to identify it. Tyuyamunite is a hydrous (water-bearing) mineral, and when it is exposed to the atmosphere it loses its water. In the process, it changes slightly and actually becomes a different mineral: metatyuyamunite. Therefore, all but the freshest specimens should be considered metatyuyamunite.

WHERE TO LOOK: Certain areas along the Huron River in Baraga County are the only places where this mineral is found.

Beach-worn unakite

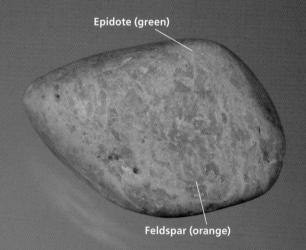

Epidote (green)

Feldspar (orange)

Unakite

HARDNESS: Varies **STREAK:** N/A

Occurrence

ENVIRONMENT: Lakeshore

WHAT TO LOOK FOR: Green and orange beach-worn pebbles

SIZE: Most unakite pebbles are golf ball-sized or smaller

COLOR: Varies; primarily yellow-green with spots of orange, pink, white, and colorless

OCCURRENCE: Common

NOTES: Named for the Unaka Mountains in Tennessee, unakite is a colorful green and orange rock easily found on most of Lake Superior's Michigan shoreline. Many casual collectors pick up the stones and use them as decorative pieces in their homes, while others collect unakite to appreciate its bright, contrasting colors. It is actually a variety of granite that is very rich in epidote, and is also known as epidotized granite. And, like the other samples of granite you'll find on the lakeshore, unakite was carried south from Canada by the glaciers that scoured the land during the past ice ages. It isn't particularly valuable, though it is often found as polished pieces in shops.

Unakite's coarsely grained appearance and characteristic colors are always enough to identify it. As mentioned above, the rock contains large amounts of yellowish-green epidote and orange or pink orthoclase feldspar. It also contains quartz, evident as white or colorless grains within the rock. It can also contain small grains of a mineral called apatite. Unlike the regular granites, it lacks dark minerals, such as micas and pyroxenes.

WHERE TO LOOK: Look on the Lake Superior shoreline along the entire length of the Upper Peninsula.

Serpentine (green) with quartz (white) veins

Polished verde antique

Serpentine

Calcite

Specimen courtesy of John Perona

Verde Antique

HARDNESS: Varies **STREAK:** N/A

Occurrence

ENVIRONMENT: Mine dumps, quarries

WHAT TO LOOK FOR: Green, greasy-feeling rock with veins of calcite or quartz

SIZE: Verde antique forms massively and can be found in any size, from pebbles to boulders

COLOR: Varies; light to dark green, white, gray

OCCURRENCE: Rare

NOTES: You may have seen verde antique used as a decorative stone in a fancy hotel lobby or as a counter-top in a century-old building—and it may have even come from Michigan. The famous material is a variety of serpentinite, which is a rock made primarily of serpentine, and it was once quarried in large amounts from the Upper Peninsula. The serpentine in verde antique is easy to identify thanks to its green color, greasy feel, and massive habit (it forms as chunks rather than crystals). Filling cracks and spaces between masses of serpentine are calcite, dolomite and sometimes quartz. The result is a white and green rock with beautiful patterns and shapes. As mentioned above, verde antique has been used for decades as a decorative architectural accent as well as a carving material.

Rough verde antique is very different in appearance from the polished pieces found in shops. While the patterns are less obvious, look for solid, chunky pieces of serpentine with the characteristic veins of white calcite or dolomite.

WHERE TO LOOK: Serpentine-rich areas, such as the old mines in the Michigan Gold District, near Ishpeming, have produced fine specimens of this rock.

Mesolite (white) in basalt

Thomsonite in basalt

Analcime on matrix

Zeolite group

HARDNESS: 3.5–5.5 **STREAK:** Colorless to white

Occurrence

ENVIRONMENT: Lakeshore, road cuts, mine dumps

WHAT TO LOOK FOR: Small, light-colored crystals, often fibrous, which fill vesicles (gas bubbles) in basalt

SIZE: Zeolites rarely occur larger than your thumbnail

COLOR: Varies depending on species; colorless, white to gray, pink, salmon to orange, and some greens

OCCURRENCE: Certain varieties are common, others are very rare

NOTES: The zeolite group contains over 50 minerals, all of which are very chemically complex, contain water, and form in basalt vesicles (gas bubbles). Many are fibrous and form as delicate needle-like crystals arranged into "sprays," or fan-shaped groupings. The majority of zeolites cannot be differentiated without laboratory analysis, so zeolite identification can be a headache for the amateur. Luckily, Michigan only has a handful of zeolites, several of which are easy to identify. Laumontite (page 143) is a generally orange, pink, or salmon-colored mineral and forms in fibrous crystals that easily dehydrate, which causes them to crumble and makes them easy to pick apart. Thomsonite (page 225) is a rare pink and green zeolite that often has "eyes," or circular formations within it. Mesolite looks very much like thomsonite, but often lacks the pink color. And analcime resembles calcite more than it does other zeolites, though it is harder. It forms as small, round colorless or white crystals. All these minerals form within basalt vesicles (gas bubbles), often filling them completely.

WHERE TO LOOK: Look in basalt formations on the Keweenaw Peninsula, particularly near the shore on the peninsula's northernmost tip.

Glossary

AGGREGATE: An accumulation or mass of crystals

ALTER: Chemical changes within a rock or mineral due to the addition of mineral solutions

AMPHIBOLE: A large group of important rock-forming minerals that commonly have a fibrous appearance

AMYGDULE: A vesicle, or gas bubble, filled with a secondary mineral

ASSOCIATED: Minerals that often occur together due to similar chemical traits

ASBESTOS: A very fibrous, flexible, silky-feeling mineral formation; it can refer to several different minerals

BAND: An easily identified layer within a mineral

BED: A large, flat mass of rock, generally sedimentary

BOTRYOIDAL: Crusts of a mineral that formed in rounded masses, resembling a bunch of grapes

BRECCIA: A coarse-grained rock composed of broken angular rock fragments solidified together

CHALCEDONY: A massive, microcrystalline variety of quartz

CLEAVAGE: The property of a mineral to break along the planes of its crystal structure, which reflects its internal crystal shape

COMPACT: Dense, tightly formed rocks or minerals

CONCENTRIC: Circular, ringed banding, resembling a bull's-eye pattern, with larger rings encompassing smaller rings

CONCHOIDAL: A rounded shape resembling a half-moon, generally referring to fracture

CRYSTAL: A solid body with a repeating atomic structure formed when an element or chemical compound solidifies

CUBIC: A box-like structure with sides of an equal size

DEHYDRATE: To lose water contained within

DRUSY: A coating of small crystals on the surface of another rock or mineral

DULL: A mineral that is poorly reflective

EARTHY: Resembling soil; dull luster and rough texture

EFFERVESCE: When a mineral placed in an acid gives off bubbles caused by the mineral dissolving

FELDSPAR: An extremely common and diverse group of light-colored minerals that are most prevalent within rocks and make up the majority of the earth's crust

FIBROUS: Fine, rod-like crystals that resemble cloth fibers

FLOAT: Metal that has been scraped up, moved and rounded by glaciers, resulting in a "nugget"

FLUORESCENCE: The property of a mineral to give off visible light when exposed to ultraviolet light radiation

FRACTURE: The way a mineral breaks or cracks when struck, often referred to in terms of shape or angles

GLASSY: A mineral with a reflectivity similar to window glass, also known as "vitreous luster"

GNEISS: A rock that has been metamorphosed so that some of its minerals are aligned in parallel bands

GRANITIC: Pertaining to granite or granite-like rocks

GRANULAR: A texture or appearance of rocks or minerals that consist of grains or particles

HEXAGONAL: A six-sided structure

HOST: A rock or mineral on or in which other rocks and minerals occur

HYDROUS: Containing water

IGNEOUS ROCK: Rock resulting from the cooling and solidification of molten rock material, such as magma or lava

IMPURITY: A foreign mineral within a host mineral that often changes properties of the host, particularly color

INCLUSION: A mineral that is encased or impressed into a host mineral

IRIDESCENCE: When a mineral exhibits a rainbow-like play of color

LAMELLAR: Minerals composed of thin parallel crystals arranged into book- or gill-like aggregates

LAVA: Molten rock that has reached the earth's surface

LUSTER: The way in which a mineral reflects light off of its surface, described by its intensity

MAGMA: Molten rock that remains deep within the earth

MASSIVE: Minerals that don't occur in individual crystals but rather as solid, compact concentrations; rocks are often described as massive; in geology, "massive" is rarely used in reference to size

MATRIX: The rock in which a mineral forms

METAMORPHIC ROCK: Rock derived from the altering of existing igneous or sedimentary rock through the forces of heat and pressure

METAMORPHOSED: A rock or mineral that has already undergone metamorphosis

MICA: A large group of minerals that occur as thin flakes arranged into layered aggregates resembling a book

MICACEOUS: Mica-like in nature; said of a mineral consisting of thin sheets

MICROCRYSTALLINE: Crystal structure too small to see with the naked eye

MINERAL: A naturally occurring chemical compound or native element that solidifies with a definite internal crystal structure

NATIVE ELEMENT: An element found naturally uncombined with any other elements, e.g. copper

NODULE: A rounded mass consisting of a mineral, generally formed within a vesicle

OCTAGONAL: An eight-sided structure

OCTAHEDRAL: A structure with eight-faces, resembling two pyramids placed base-to-base

OPAQUE: Material that lets no light through

ORE: Rocks or minerals from which metals can be extracted

OXIDATION: The process of a metal or mineral combining with oxygen, which can produce new colors or minerals

PEARLY: A mineral with reflectivity resembling that of a pearl

PEGMATITE: Portions of granite formations, often rich with water, that cool slowly, allowing the minerals within the magma to more fully crystallize, resulting in large and sometimes rare crystals

PHENOCRYST: A crystal embedded within igneous rock that solidified before the rest of the surrounding rock, thus retaining its true crystal shape

PLACER: Deposit of sand containing dense, heavy mineral grains at the bottom of a river or a lake

PORPHYRY: An igneous rock containing many phenocrysts

PRISMATIC: Crystals with a length greater than their width

PSEUDOMORPH: When one mineral replaces another but retains the outward appearance of the initial mineral

PYRAMIDAL: Crystals resembling a pyramid with four or more total faces

PYROXENE: A group of dark, rock-building minerals that make up many dark-colored rocks like basalt or gabbro

RADIATING: Crystal aggregates growing outward from a central point, resembling the shape of a paper fan

RHOMBOHEDRON: A six-sided shape resembling a tilted or leaning cube

ROCK: A massive aggregate of mineral grains

ROCK-FORMING: Refers to a mineral important in rock creation

SCHIST: A rock that has been metamorphosed so that most of its minerals have been concentrated and arranged into parallel layers

SECONDARY: A rock or mineral that formed later than the rock surrounding it

SEDIMENT: Fine particles of rocks or minerals deposited by water or wind, e.g. sand

SEDIMENTARY ROCK: Rock derived from sediment being cemented together

SILICA: Silicon dioxide, more commonly known as quartz

SMELTING: Processing a rock or mineral, usually by melting, in order to separate metals

SPECIFIC GRAVITY: The ratio of the density of a given solid or liquid to the density of water when the same amount of each is used, e.g. the specific gravity of galena is approximately 7.5, meaning that a sample of galena is about 7.5 times heavier than the same amount of water

SPECIMEN: A sample of a rock or mineral

STALACTITIC: Resembling a stalactite, which is a cone-shaped mineral deposit; sometimes described as carrot-shaped

STRIATED: Parallel grooves in the surface of a mineral

TABULAR: A crystal structure in which one dimension is notably shorter than the others, resulting in flat, plate-like shapes

TARNISH: A thin coating on the surface of a metal, often differently colored than the metal itself (see *oxidation*)

TRANSLUCENT: A material that lets some light through

TRANSPARENT: A material that lets enough light through as to be able to see what lies on the other side

TWIN: An intergrowth of two or more crystals

VEIN: A mineral, particularly a metal, that has filled a crack or similar opening in a host rock or mineral

VESICLE/VESICULAR: A cavity created in an igneous rock by a gas bubble trapped when the rock solidified; a rock containing vesicles is said to be vesicular

VUG: A small cavity within a rock or mineral that can become lined with different mineral crystals

WAXY: A mineral with a reflectivity resembling that of wax, such as a candle

Michigan Rock Shops and Museums

Upper Peninsula

A.E. SEAMAN MINERAL MUSEUM
Michigan Technological University
1404 Sharon Ave.
Houghton, MI 44931
(906) 487-2572
www.museum.mtu.edu

CLIFFS SHAFT MINE MUSEUM
501 W. Euclid Street,
Ishpeming, MI 49849
(906) 485-1882

GITCHE GUMEE AGATE AND HISTORY MUSEUM
P.O. Box 308,
East 21739 Brazel Street
Grand Marais, MI 49839
(906) 494-2590

IRON MOUNTAIN IRON MINE (rock shop and mine tours)
9 miles east of Iron Mountain on US-2, MI 49801
(906) 563-8077

KEWEENAW GEM AND GIFT, INC. (rock shop)
1007 West Memorial
Houghton, MI 49931
(906) 482-8447

NATURE'S PICKS ROCK SHOP
600 Cloverland Avenue
Ironwood, MI 49938
(906) 932-7340

PROSPECTOR'S PARADISE (rock shop)
3 miles north of Calumet
US-41, Keweenaw Peninsula, MI 59691
(906) 337-6889

QUINCY MINE HOIST (rock shop and mine tours)
49750 US Highway 41
Hancock, MI 49930
(906) 482-3101

RED METAL MINERALS (rock shop and mine tours)
202 Ontonagon St.
Ontonagon, MI 49953
(906) 884-6618

ROCK KNOCKER'S ROCK SHOP
490 North Street (US-41)
Ishpeming, MI 49849
(906) 485-5595

SWEDES (rock shop)
P.O. Box 66, US-41
Copper Harbor, MI 49918
(906) 289-4596

Lower Peninsula

CRANBROOK INSTITUTE OF SCIENCE (museum)
39221 Woodward Ave.
Bloomfield Hills, MI 48303-0801
www.cranbrook.edu

**THE HALL OF IDEAS AT THE MIDLAND CENTER
FOR THE ARTS** (museum)
1801 W. Saint Andrews Rd.
Midland, MI 48640
(989) 631-5930
www.mcfta.org

MICHIGAN HISTORICAL MUSEUM
717 W. Allegan St.
Lansing, MI 48918
(517) 373-3559

THE ROCK SHOPPE
6275 Gotfredson Rd.
Plymouth, MI 48170
(734) 455-5560
http://rock-shoppe.com

UNIVERSITY OF MICHIGAN-DEARBORN (museum)
Department of Natural Sciences
125 Science Building
4901 Evergreen Rd.
Dearborn, MI 48128
(313) 436-9129

Bibliography and Recommended Reading

Books about Michigan Minerals

Brzys, Karen. *Understanding and Finding Agates*. Hancock: Book Concern Printers, 2004.

Carlson, Michael. *The Beauty of Banded Agates*. Edina: Fortification Press, 2002.

Heinrich, E. W., et al. *Mineralogy of Michigan*. Houghton: Michigan Technical University, 2004.

Marshall, John. *The "Other" Lake Superior Agates*. Beaverton: Llao Rock Publications, 2003.

Pabian, Roger, et al. *Agates: Treasures of the Earth*. Buffalo: Firefly Books Limited, 2006.

Robinson, Susan. *Is This an Agate?* Hancock: Book Concern Printers, 2001.

Vicary, Joel. *Over 100 Collecting Locations in Lower Michigan*. Self-published, 2007.

Zeitner, June Culp. *Midwest Gem, Fossil and Mineral Trails of the Great Lakes States*. Baldwin Park: Gem Guides Book Company, 1999.

General Reading

Bates, Robert L., editor. *Dictionary of Geological Terms, 3rd Edition*. New York: Anchor Books, 1984.

Bonewitz, Ronald Louis. *Smithsonian Rock and Gem*. New York: DK Publishing, 2005.

Chesteman, Charles W. *The Audubon Society Field Guide to North American Rocks and Minerals*. New York: Knopf, 1979.

Johnsen, Ole. *Minerals of the World*. New Jersey: Princeton University Press, 2004.

Mottana, Annibale, et al. *Simon and Schuster's Guide to Rocks and Minerals*. New York: Simon and Schuster, 1978.

Pellant, Chris. *Rocks and Minerals*. New York: Dorling Kindersley Publishing, 2002.

Pough, Frederick H. *Rocks and Minerals*. Boston: Houghton Mifflin, 1988.

Index

About the Authors

Dan R. Lynch has a degree in graphic design with emphasis on photography from the University of Minnesota Duluth. But before his love of the arts came a passion for rocks and minerals, developed during his lifetime growing up in his parents' rock shop. Combining the two aspects of his life seemed a natural choice and he enjoys both writing about and taking photographs of minerals. Working with his father, Bob Lynch, a respected veteran of the rock-collecting community, Dan spearheads their series of rock and mineral field guides—definitive guidebooks useful for rock hounds of any skill level. Dan takes special care to ensure that his photographs complement the text and represent each rock or mineral exactly as you'll find them. Encouraged by his wife, Julie, he works as a writer and photographer.

Bob Lynch is a lapidary and jeweler working in Two Harbors, Minnesota. He has been working with rocks and minerals since 1973, when he desired more variation in gemstones for his work with jewelry. When he moved from Douglas, Arizona, to Two Harbors in 1982, his eyes were opened to Lake Superior's entirely new world of minerals. In 1992, Bob and his wife Nancy, another jeweler, acquired Agate City Rock Shop, a family business founded by Nancy's grandfather, Art Rafn, in 1962. Since the shop's revitalization, Bob has made a name for himself as a highly acclaimed agate polisher. Now, the two jewelers keep Agate City Rocks and Gifts open year-round and are the leading source for Lake Superior agates, with more on display and for sale than any other shop in the country.

Notes

"A powerful account of Teege's struggle for resolution and redemption, the book [is] itself a therapeutic working-through of her history, as well as a meditation on family."   —*The Independent* (UK)

"Courageous. . . . the memoir invites rereading to fully absorb Teege's painful search for answers, for a sense of identity and belonging and for inner peace. Readers won't help but feel for her. Teege discovers, however, that history's shattering truths have the potential to make us more whole."   —*Seattle Times*

"[Teege's] message is an important one—that we have the power to decide who we are."   —*Seattle Weekly*

"In honest, direct, and absorbing prose, Teege and coauthor Nikola Sellmair confront highly personal repercussions of the Holocaust. . . . The book's real triumph is in its nuanced, universally appealing portrait of an individual searching for her place in the world. Just as Teege's chance encounter with a library book led her to question the fundamental assumptions of her life, so too the reader. . . will be forced to reconsider the wide-ranging impact of past injustices on present-day relationships." —**The Jewish Book Council**

"A discomfiting but clear-eyed journey of self-discovery and identity reconciliation that first-time author Teege relates with admirable straightforwardness and equanimity."   —*In These Times*

"The alternating narrative between Teege and co-author Sellmair offers a refreshing and ultimately impartial analysis. Teege's heartfelt commentary and Sellmair's objective narrative produce a layer of balanced interpretation and insight."   —*New York Journal of Books*

"Teege's story is at times heart wrenching, and yet, full of her own stark honesty and surprising wisdom as she ponders the impacts of one's family history."   —*Manhattan Book Review*

"Jennifer Teege has a fascinating story."
                              —*Washington Independent Review of Books*

"Teege's story is one of questions as much as answers. Her honest self-examination makes for a provocative, unpredictable story of an understanding still in progress."   —*Columbus Dispatch*

MY GRANDFATHER
WOULD HAVE SHOT ME

MY GRANDFATHER
WOULD HAVE SHOT ME

A BLACK WOMAN DISCOVERS
HER FAMILY'S NAZI PAST

JENNIFER TEEGE
AND NIKOLA SELLMAIR

TRANSLATED BY CAROLIN SOMMER

THE EXPERIMENT

NEW YORK

My Grandfather Would Have Shot Me: *A Black Woman Discovers Her Family's Nazi Past*

Originally published under the title *Amon: Mein Grossvater hätte mich erschossen*
First published in North America by The Experiment, LLC, in 2015
Copyright © 2013, 2015 by Rowohlt Verlag GmbH, Reinbek bei Hamburg
Copyright © 2013, 2015 by Jennifer Teege and Nikola Sellmair
Translation copyright © 2015 by Carolin Sommer
New material copyright © 2016 by The Experiment
The translation of this work was supported by a grant from the Goethe-Institut
which is funded by the German Ministry of Foreign Affairs.

All photographs are courtesy of the author or in the public domain, except for the following, which are used with permission: ullstein bild, Berlin: 8 (AP), 57 (imagebroker.net/Petr Svarc); Nikola Sellmair/stern/Picture Press, Hamburg: 45; Sueddeutsche Zeitung Photo, Munich: 28 (Teutopress); Yad Vashem Photo Archive, Jerusalem: 33, 39, 51, 53, 63, 91 (Emil Dobel); Diane Vincent, Berlin: 194, 196

Some of the names of people, and identifying details, have been changed to protect the privacy of individuals.

Many of the designations used by manufacturers and sellers to distinguish their products are claimed as trademarks. Where those designations appear in this book and The Experiment was aware of a trademark claim, the designations have been capitalized.

The Experiment, LLC, 220 East 23rd Street, Suite 600, New York, NY 10010-4658
www.theexperimentpublishing.com

The Experiment's books are available at special discounts when purchased in bulk for premiums and sales promotions as well as for fund-raising or educational use. For details, contact info@ theexperimentpublishing.com.

Library of Congress Cataloging-in-Publication Data

Teege, Jennifer, 1970-
[Amon. English]
My grandfather would have shot me : a Black woman discovers her family's Nazi past / Jennifer Teege, Nikola Sellmair.
     pages cm
"Originally published under the title Amon : mein Grossvater hätte mich erschossen"--Title page verso.
Includes bibliographical references.
ISBN 978-1-61519-253-3 (cloth) — ISBN 978-1-61519-254-0 (ebook)
1. Teege, Jennifer, 1970- 2. Teege, Jennifer, 1970---Family. 3. Grandchildren of war criminals—
Germany—Biography. 4. Racially mixed people—Germany—Biography. 5. Göth, Amon,
1908-1946--Family. 6. Nazis—Family relationships. 7. Concentration camp commandants—Family
relationships. 8. Plaszów (Concentration camp) 9. Teege, Jennifer, 1970—Travel—Poland. 10. Teege,
Jennifer, 1970—Homes and haunts—Israel. I. Sellmair, Nikola, 1971- II. Title.
     CT1098.T33A3 2015
     929.20943—dc23
                              2014046242

ISBN 978-1-61519-308-0
Ebook ISBN 978-1-61519-254-0

Cover design by Sarah Smith | Text design by Pauline Neuwirth, Neuwirth & Associates, Inc. Author photograph © Marcelo Hernandez | Cover photograph of Kraków-Płaszów concentration camp © Photo Archive, Yad Vashem, Jerusalem

Manufactured in the United States of America

First paperback printing April 2016
10 9 8 7 6 5 4

FOR Y.

CONTENTS

THE DISCOVERY

IT IS THE LOOK on the woman's face that seems familiar. I'm standing in the central library in Hamburg, and in my hands I'm holding a red book that I've just picked up from the shelf. The spine reads: *I Have to Love My Father, Don't I?* On the front cover is a small black-and-white photograph of a middle-aged woman. She looks deep in thought, and there is something strained and joyless about her. The corners of her mouth are turned down; she looks unhappy.

I glance quickly at the subtitle: *The Life Story of Monika Goeth, Daughter of the Concentration Camp Commandant from "Schindler's List."* Monika Goeth! I know that name; it's my mother's name. My mother, who put me in an orphanage when I was little and whom I haven't seen in many years.

I was also called Goeth once. I was born with that name, wrote "Jennifer Goeth" on my first schoolbooks. It was my name until after I was adopted, when I took on the surname of my adoptive parents. I was seven years old.

Why is my mother's name on this book? I am staring at the cover. In the background, behind the black-and-white photo of the woman, is a shadowy picture of a man with his mouth open and a rifle in his hands. That must be the concentration camp commandant.

I open the book and start leafing through its pages, slowly at first, then faster and faster. It contains not only text but lots of photos, too. The people in the pictures—haven't I seen them somewhere before? One is a tall, young woman with dark hair; she reminds me of my mother. Another is an older woman in a flowery summer dress, sitting in the English Garden in Munich. I don't have many pictures of my grandmother, but I know each of them very well. In one of them she is wearing the exact same dress as this woman. The caption under the photo says *Ruth Irene Goeth.* That was my grandmother's name.

Is this my family? Are these pictures of my mother and my grandmother? Surely not, that would be absurd: It can't be that there is a book about my family and I know nothing about it!

I quickly skim through the rest of the book. Right at the back, on the last page, I find a biography, and it begins like this: "Monika Goeth, born in Bad Toelz in 1945." I know these dates; they are on my adoption papers. And here they are, in black and white. It really is my mother. This book *is* about my family.

I snap the book shut. It is quiet. Somewhere in the reading room someone is coughing. I need to get out of here, quickly; I need to be alone with this book. Clutching it close to me like a precious treasure, I just barely manage to walk down the stairs and through the checkout. I don't take in the librarian's face as she hands the book back to me. I walk out onto the expansive square in front of the library. My knees buckle. I lie down on a bench and close my eyes. Traffic rushes past me.

My car is parked nearby, but I can't drive now. A couple of times I sit up and consider reading on, but I am dreading it. I want to read the book at home, in peace and quiet, cover to cover.

It is a warm, sunny August day, but my hands are as cold as ice. I call my husband. "You have to come and get me; I have found a book. About my mother and my family."

Why did my mother never tell me? Do I mean that little to her, still? Who is this Amon Goeth? What exactly did he do? Why do I know nothing about him? What was the story of *Schindler's List* again? And what about the people I've heard referred to as Schindler Jews?

It has been a long time since I've seen the film. I remember that it came out in the middle of the 1990s, while I was studying in Israel. Everybody was talking about Steven Spielberg's Holocaust movie. I didn't watch it until later, on Israeli TV, alone in my room in my shared flat, in Rehov Engel—Engel Street—in Tel Aviv. I recall that I was touched by the film, but that I thought the end was a bit kitschy, too Hollywood.

Schindler's List was just a film to me; it didn't have anything to do with me personally.

Why has nobody told me the truth? Has everybody been lying to me for all these years?

CHAPTER 1

ME, GRANDDAUGHTER OF A MASS MURDERER

In Germany, the Holocaust is family history.
—RAUL HILBERG

WAS BORN ON JUNE 29, 1970, the daughter of Monika Goeth and a Nigerian father. When I was four weeks old, my mother took me to a Catholic orphanage and put me in the care of the nuns.

At three, I was taken in by a foster family, who then adopted me when I was seven. My skin is black, while that of my adoptive parents and two brothers is white. Everybody could see that I was not their biological child, but my adoptive parents always reassured me that they loved me just as much as their own children. They took me and my brothers to playgroups and Gymboree classes. As a child, I still saw my biological mother and grandmother, but we lost contact as I grew older. I was 21 when I last saw my mother.

Then, at age 38, I found *the book*. Why on earth did I pick it up off the shelf, one among hundreds of thousands of books? Is there such a thing as fate?

The day had begun just as usual. My husband had gone to work; I had taken my sons to preschool and then gone into town to visit the library. I go there often. I like the concentrated silence, the quiet footsteps, the rustling noise of turning pages, the reading visitors hunched over their books. I was looking for something about depression in the psychology department. There, at hip level, between Erich Fromm's *The Art of Loving* and a book with the vague title *The Power Lies in the Crisis,* was the book with the red cover. I had never heard of the author, Matthias Kessler, but the title sounded interesting: *I Have to Love My Father, Don't I?* So I took the book from the shelf.

When my husband Goetz comes to pick me up, he finds me lying on the bench in front of the library. He sits down beside me, examines the book, and starts leafing through its pages. I snatch it back from him. I don't want him to read it first because I've realized that the book is meant for me, the key to my family history, to my life. The key I've been looking for all these years.

For my whole life I had felt that there was something wrong with me: behind my sadness, my depression. But I could never quite put my finger on what was so fundamentally wrong.

Goetz takes my hand and we walk over to his car. I hardly say a word on the way home. He takes the rest of the day off and looks after our two sons.

I collapse onto our bed and read and read, to the very last page. It is dark when I close the book. Then I sit down at my computer and spend the whole night online, reading everything I can find about Amon Goeth. I feel like I have entered a chamber of horrors. I read about his decimation of the Polish ghettoes, his sadistic murders, the dogs he trained to tear humans apart. It is only now that I realize the magnitude of the crimes Amon Goeth committed. Himmler, Goebbels, Goering—I know who they are. But what exactly Amon Goeth had done, I'd had no idea. Slowly I begin to grasp that the Amon Goeth in the film *Schindler's List* is not a fictional character, but a person who actually existed in flesh and blood. A man who killed people by the dozens and, what is more, who enjoyed it. My grandfather. I am the granddaughter of a mass murderer.

◾ ◾ ◾

Jennifer Teege has a deep, warm voice with a hint of a Munich accent, slightly rolling her "R"s. Her face is bright, she doesn't wear make-up; her naturally frizzy hair is tamed into long black curls. Tight fitted pants hug her long, thin legs. When she enters a room she turns heads, and men's eyes follow her around. She walks upright, with a firm, determined step.

Her friends describe Jennifer as a confident woman, inquisitive and full of adventure. A college friend says, "If she heard about an exciting country, she'd say, 'I've not been there, I'll go and visit!' And off she'd go—to Egypt, Laos, Vietnam, and Mozambique."

But when she talks about her family history, her hands tremble and she begins to cry.

The moment when Jennifer found the book with the library code Mcm O GOET#KESS is the moment that cut her life in two, into a before and an after: A *before*, when she lived without knowledge of her family's past, and an *after*, living with that knowledge.

The whole world knows her grandfather's story: In Steven Spielberg's film *Schindler's List*, the cruel concentration camp commandant Amon Goeth is Oskar Schindler's drinking buddy and adversary: Two men born in the same year, one a murderer of Jews, the other their savior. One particular scene has stuck in the collective memory: Amon Goeth shooting prisoners from his balcony, his personal form of morning exercise.

As commandant of the Płaszów concentration camp, Amon Goeth was responsible for the deaths of thousands of people. In 1946, he was hanged in Krakow; his ashes were thrown in the Vistula River. Goeth's lover Ruth Irene, Jennifer Teege's beloved grandmother, denied his crimes ever after. In 1983 she killed herself with an overdose of sleeping pills.

Oskar Schindler's original list, discovered in the attic of a house in Hildesheim in 1999; in the background, a photograph of Oskar Schindler (center)

That is Jennifer Teege's personal German history: her grandfather a Nazi criminal, her grandmother a follower, her mother raised in the leaden silence of the post-war period. That is her family, those are the roots that she, the adopted child, had always been looking for. But what about *her*, where does all that leave her?

◼ ◼ ◼

MY DISCOVERY LEADS ME TO QUESTION everything that had been central to my life: my close relationships with my adoptive brothers and with my friends in Israel, my marriage, my two sons. Has my whole life been a lie? I feel like I have been traveling under a false name, like I have betrayed everyone, when really it is I who was betrayed. I was the one who was cheated—out of my history, my childhood, my identity.

I no longer know to whom I belong: my adoptive family, or the Goeth family? When I was seven, after the adoption, giving up the name Goeth had seemed easy to do. A document was drawn up. My adoptive parents asked if I was OK with changing my name. I said yes. I didn't venture to ask about my biological mother after that; I wanted a normal family at last. But now it seems I have no choice in the matter: I am a Goeth.

During my Internet research on Amon Goeth, I also learn of a TV program on the culture channel "Arte." An American film-maker has documented a meeting between my mother and Helen Rosenzweig, a former concentration-camp inmate and maid in my grandfather's mansion. As it happens, the film is going to be shown for the first time on German television tomorrow.

First the book, now this film—it's too much, everything is happening too quickly.

My husband and I watch the film together. Right at the beginning, my mother appears. I lean toward the television; I want to see everything clearly: *What does she look like, how does she move, how does she talk? Am I like her?* She has dyed her hair strawberry blonde; she looks haggard. I like the way she expresses herself. When I was a child, she was just my mother to me. Children don't register whether somebody is simple or educated. Now I realize: *My mother is an intelligent woman; she is saying interesting things.*

The documentary also shows a key scene from *Schindler's List*, where the Jewish forewoman explains to the newly appointed commandant Amon Goeth that the barracks have not been planned correctly—so Goeth, played by Ralph Fiennes, has the woman shot. She manages to say, "Herr Commandant, I am only doing my job." And Fiennes, as Goeth, replies, "And I'm doing mine."

My memories of the film are beginning to come back to me. That scene had shocked me, since it shows so clearly what is so difficult to imagine: In the camp, there are no limits and no inhibitions. Common sense and humanity have been abolished.

But what can I, with my dark skin and friends all over the world, have to do with such a grandfather? Was it he who destroyed my family? Did he cast his shadow first on my mother and then on me? Can it be that a dead man still wields power over the living? Is the depression that has plagued me for so long connected to my origins? I lived and studied in Israel for five years—was that chance or fate? Will I have to behave differently toward my Israeli friends, now that I know? *My grandfather murdered your relatives.*

I am dreaming: I am swimming in a dark lake, the water as thick as tar. Suddenly corpses appear all around me: spindly figures, skeletons almost, that have had everything humane taken from them.

Why did my mother not think it necessary to inform me about my origins? Why did she tell others these things that I, too, absolutely needed to know? She never told me the truth. But I need the truth. I am reminded of Theodor W. Adorno's famous phrase, *There is no right life in the wrong one.* He meant it differently when he said it, but it seems to apply perfectly to my life now.

Ours was a difficult relationship: We met only sporadically, but she is still my mother. The book about Monika Goeth mentions the year 1970, the year of my birth. There is not a word about me; my mother pretends I don't exist.

Again and again I look at the picture in the book where she looks just as I remember her from my childhood. Deep inside my head, the drawers of my memory are opening one by one. My entire childhood comes up to the surface, all the feelings from my time in the orphanage, the loneliness and the despair.

I feel helpless again, like a small, disappointed child, and I am losing my grip on life.

All I want to do is sleep; often I stay in bed until midday. Everything feels like too much effort: having to get up, to talk. Even brushing my teeth is a struggle. The answering machine is permanently switched on, but I never manage to return any calls. I stop seeing my friends, and I turn down all invitations. What could I possibly talk or laugh about? It feels as if there is a glass wall between my family and me. How can I explain to them what I am going through, when I myself don't understand what is happening to me?

Suddenly I can no longer bear people drinking beer near me. The smell of beer alone is enough to make me sick: It reminds me of my mother's first husband. When he was drunk, which he usually was, he would beat my mother.

For two weeks after discovering the book, I hardly leave the house. Sometimes I manage to pull on a pair of jeans instead of the usual sweatpants, but I'm soon overwhelmed by crushing tiredness and wonder why I have bothered to shower and get dressed when I am not going out anyway.

My husband does his best to look after our children. He gets the groceries on the weekends, fills up the freezer and cooks meals in advance. I don't want to be a bad mother who just leaves her sons to watch TV in the afternoons. Instead I go online and order some Legos for them; it will keep my children busy for a few hours while I get some rest.

Finally I try once again to go out, to look after my family, but I falter at the smallest hurdles. In the supermarket, the crowds make me nervous. Baffled, I find myself staring at the different types of coffee on the shelf. Surely I have much more urgent business to do at the post office? So I go to the post office instead, but once there I find that the line is too long, and I hurry back to the supermarket, back to the coffee shelf. I remember that I had actually wanted some milk and bread. But much more important is lunch—now where am I going to get that? It is getting late, and I need to go and fetch the children from preschool soon. The pressure is rising, my head is my prison. Once again, I've gotten nothing done.

I never had a real mother myself, and so I've tried to give my children everything I never had, but now I'm deserting them. I make sandwiches for them and heat up TV dinners. Simple, functional things; nothing more. My older son Claudius craves my company. At bedtime, he wants lots of cuddles and talks to me, fast and nonstop, so as not to allow any gaps in the conversation where I might turn away again. I try to concentrate on what he is saying, but I can't. I nod my head every now and then to pretend I am listening. I would love to just pull the blanket over my head.

Why didn't I discover that I am the granddaughter of some great Nobel Prize winner?

■　■　■

Anybody who is related to Joseph Goebbels, Heinrich Himmler, Hermann Goering or an Amon Goeth is compelled to deal with their family history. But what about all the others, the many unnamed followers and accomplices?

In his research study *Grandpa Wasn't a Nazi*, social psychologist Harald Welzer came to the following conclusion: The generation of grandchildren, today's 30- to 50-year olds, tend to know the facts about the Holocaust and often reject the Nazi ideology even more strongly than the previous generation. Their critical eye, however, is only directed at political issues—not at private affairs. The grandchildren in particular sugarcoat the role their ancestors played: Two-thirds of those questioned even stylized their forebears into heroes of the Resistance or victims of the Nazi regime themselves.

Many have no idea what their own grandfathers were really up to. To them, the Holocaust is a history class, the victims' story memorialized in films and on TV; they don't look at it as the history of their own family, their own personal history. So many innocent grandfathers, so many suppressed family secrets. Soon the last witnesses will be gone, and it will be too late for the grandchildren to ask questions.

■　■　■

Amon Goeth in 1945, after his arrest by the Americans

WHEN I LOOKED IN THE MIRROR as a child, it was obvious that I was different: My skin was dark, my hair frizzy. All around me there were only short, blond people: my adoptive parents and my two adoptive brothers. By contrast, I was a tall child with skinny legs and black hair. Back then, in the seventies, I was the only black child in Waldtrudering, the tranquil, leafy neighborhood in Munich where I lived with my adoptive family. At school we sometimes sang the nursery rhyme "Zehn Kleine Negerlein" (Ten little negroes)—and I hoped that nobody would turn around and look at me, that nobody would realize I didn't really belong.

Since that day in the library I have been looking in the mirror again, but now I'm looking for similarities. I'm terrified of belonging now, of belonging to the Goeths: The lines between my nose and my mouth are just like my mother's and my grandfather's. A thought flashes across my mind: I must do something about these lines, must have them botoxed, lasered, lifted!

I am tall like my mother, like my grandfather. When Amon Goeth was hanged after the war, the executioner had to shorten the rope twice; he had underestimated how tall Goeth was.

My grandfather's execution was recorded on film, so there would be proof that he really was dead. It is not until the third attempt that he ends up swinging on the rope with a broken neck. When I watched the film, I didn't know whether to laugh or cry.

My grandfather was a psychopath, a sadist. He embodies everything that I condemn. What kind of person takes pleasure in tormenting and killing others, in inventing different ways of doing so? During my research I find no explanation for why he turned out to be like that. He had seemed normal as a child.

On the matter of blood: What did I inherit from him? Does his violent temper manifest itself in me and my children? In the book about my mother, I read that she spent some time as a patient in a psychiatric hospital. The book also mentions that my grandmother kept small pink pills called Prolixin in her bathroom cabinet. I learn that it is an antipsychotic drug used to treat depression, anxiety disorders, and hallucinations.

I no longer trust myself: Am I going mad, too? Am I already mad? At night I am plagued by terrible nightmares. In one I am in a psychiatric hospital, running through the corridors trying to escape. I jump out of a window into a courtyard and am free at last.

I make an appointment with the therapist who used to treat my depression when I was still living in Munich, and I travel to see her.

Before the appointment, I have some time, so I decide to take a detour to Hasenbergl, the poor neighborhood of Munich where my biological mother used to live. Sometimes she would come and take me home for the weekend. It still looks just as it did then, only the

15

façades of the buildings are more colorful now, the dirty gray and beige walls have been painted yellow and orange. The balconies are strung with laundry; the lawns are littered with trash. I am standing outside the apartment block where my mother used to live when someone comes out of the building and holds the door open for me. I walk up and down the different corridors, trying to remember which floor she used to live on. I think it was the second floor, where I feel a familiar sense of trepidation. I was never happy here.

Next I take the subway to Schwabing, Munich's hip and trendy neighborhood. I walk past the beautiful old church at Josephsplatz and turn into Schwind Strasse. My grandmother's former apartment is in an old, prewar building with a chestnut tree in the courtyard. The front door is open; I climb up the wooden stairs right to the top floor. My grandmother was the first person to give me solace and comfort, but the book has taken away any positive feelings I had for her. Who was this woman who spent a year and a half living with my grandfather in a villa next to the Płaszów concentration camp?

I also have an appointment with Child Protective Services. The social worker is very nice and does her best to help me, but I am only allowed to read part of my file. I ask her if there are any notes in the records indicating that I had any mental disorders as a child.

The thing is, there are things I don't know about myself that others take for granted. If a doctor asked me what illnesses ran in my family, I could never answer that question. Nor do I know if I had a pacifier as a baby, what songs I used to sing along to, or what my first cuddly toy was. I didn't have a mother to ask about these things.

No, says the lady from the agency, *there's no mention of any strange behavior.* I was a normally developed, happy child.

I just about manage to make it to my appointment with my former therapist in time. What I want to know from her is, *What was her diagnosis back then? Was I really just depressed, or did I suffer from a more severe illness? Do I appear lucid to her now?* She reassures me that it really was just that, depression, and that she never diagnosed anything else. She does admit, however, that she is out of her depth with my current situation and so refers me to her Munich-based colleague, Peter Bruendl.

※　※　※

Psychoanalyst Peter Bruendl remembers Jennifer Teege well. "There was this confident, tall, beautiful woman who asked very specific questions: How do I deal with my history?" Bruendl, an elderly gentleman with a gray beard and in a black suit, has treated several grandchildren of Nazi criminals in his Munich practice. He says, "Violence and brutalization have a deep impact on the generations that follow. What makes them ill, however, is not the crimes themselves but the silence that surrounds them. There is an unholy conspiracy of silence in perpetrator families, often spanning generations."

Guilt cannot be inherited, but *feelings* of guilt can. The children of perpetrators subconsciously pass their fears and feelings of shame and guilt on to their children, says Bruendl. This affects more families in Germany than one might think.

"Jennifer Teege's case was exceptional because she suffered a double trauma," Bruendl says, "first being given up for adoption and then the discovery of her family history."

He goes on, "Frau Teege's experience is heartbreaking. Even her conception was a provocation: Her mother, Monika Goeth, had a child with a Nigerian man. In Munich in the early 1970s, this was most unusual. And for the daughter of a concentration camp commandant it was unheard of."

Often, the grandchildren of Nazis come to him for totally different reasons, says Bruendl: depression, unwanted childlessness, eating disorders, or fear of failure at work. Bruendl encourages them to research their past and to tear down their family's web of lies. "It is only then that they can live their own lives, their own, authentic lives."

■ ■ ■

HERR BRUENDL REFERS ME to the Institute for Psychiatry at the University Hospital in Hamburg. Yet the expert he recommends is not available right away, and with every day that I have to wait I grow more and more desperate. I know that I need professional help and that everybody else is unsure how to deal with me. Sometimes I lose my temper and shout at Goetz and the children. I cannot pull myself together, cannot hold myself together anymore.

One morning, when I start to cry just after getting up, my sons ask me, *Mommy, what's wrong?*

Nothing, I answer—and then take myself to the psychiatric emergency room at the local University Hospital. The doctor on duty prescribes me antidepressants, and I take them straight away.

During the following weeks I feel like I have been superficially restored to my usual self. Then, at last, I have my appointment with the recommended therapist. His professor's office, where we meet, may be austere, but he understands my internal suffering. When he hears my story, he cries with me, and I feel in safe hands with him. My therapist will not cry with me again, but he will take good care of me in the coming months.

I start running again. I have always enjoyed my own company—traveling alone, running alone. One of my favorite places to run is a small wooded area in Hamburg. I start in the cool shade of the trees, and continue along fields and past paddocks with horses in them. I run through community gardens with garden gnomes among the flowerbeds: This pointedly idyllic world has something touching about it, and afterward my head feels clearer.

My adoptive family still don't know; I will tell them at Christmas, when we all come together at their house in Munich.

My Christmas present to everyone is a copy of the book about my mother, plus the only biography ever written of Amon Goeth. It is a hefty volume, authored by a Viennese historian.

My adoptive parents, Inge and Gerhard—I cannot call them Mama and Papa anymore—are surprised and shocked. When I discovered the book, I had first suspected that they knew everything about my biological family but that they had wanted to protect me, that they, too, had betrayed me. Yet I soon realized that they would never have kept anything so fundamental hidden from me. Their reaction now tells me that I was right: They knew nothing either.

My adoptive parents have always struggled to talk about feelings, and now it is no different. They escape into academic details: *Amon Goeth's biography has no footnotes,* my father complains, and then asks, *Does the number of dead correspond with other sources?* My life has been turned upside down, and my adoptive parents are discussing footnotes! My brothers, on the other hand, understand straight away what this book means to me.

    ▪   ▪   ▪

Jennifer Teege's adoptive mother clearly remembers the moment when Jennifer was sitting on the sofa that Christmas Eve, struggling for words. "Jenny announced that there was something important that she wanted to talk about. But then she just looked at us and started welling up. I sensed that something bad had happened." When Inge Sieber had heard the whole story, she didn't know how to deal with it at first. "My husband and I felt as if somebody had pulled the rug out from under our feet."

Jennifer Teege's adoptive brother Matthias couldn't sleep that Christmas night. "I was worried about how the discovery would affect Jenny. With this book, a different world had opened up for her. She had found her other self; she had seen where she had come from. She had spent a lot of time dealing with who her grandfather was, and even more so with the women in her family, her mother and her grandmother."

Suddenly, Jennifer regarded herself less as the daughter of her adoptive parents; instead she saw herself as part of her natural family. Matthias thinks that this upset his parents very much.

He himself was very worried about his sister: "She was so gloomy, so depressed, like I had never seen her before. I had always thought she was so strong. She was always the boldest, the most daring of us three children."

■　■　■

IN THE FOLLOWING MONTHS, my brother Matthias turns into my most important confidant, after my husband. In his research, he digs up more and more details about the Goeth family.

My Israeli friends Noa and Anat keep sending me emails: *Jenny, where are you, what's going on?* I have neither the strength nor the words to reply. I don't want to hurt my friends' feelings. I can't remember exactly where they lost their relatives during the Holocaust. I'll have to ask them—but what if they say, *In Płaszów?*

The thing is, Amon Goeth's victims are not abstract figures to me, not just an anonymous crowd. When I think of them, I see the faces of the elderly people I met during my studies in Israel at the

Goethe Institute. They were Holocaust survivors who wanted to hear and speak German again, the language of their old home country. Some could not see very well, so I read German newspapers and novels to them. On their forearms I could see their prisoners' numbers from the camps tattooed on their skin. For the first time my German nationality felt wrong, like something I had to apologize for. Luckily I was well camouflaged—thanks to my dark skin, nobody suspected that I was German.

How would those survivors have reacted if they had known that I was Amon Goeth's granddaughter? Maybe they would have wanted nothing to do with me. Maybe they would have seen him in me.

My husband tells me, *Go and find your mother's address and confront her with your anger and your questions. And tell your friends in Israel why you haven't been in touch.*

Not yet, I reply. I need to think first. And I have to visit some graves. In Krakow.

CHAPTER 2

MASTER OF THE PŁASZÓW CONCENTRATION CAMP:

MY GRANDFATHER AMON GOETH

If he liked you, you lived, if he didn't, you were dead.
—MIETEK PEMPER, FORMER ASSISTANT TO AMON GOETH

CAREFULLY I PLACE ONE FOOT in front of the other. The floor beneath me sways; the rotten wood creaks and yields under the pressure of each step. It is cold and damp in here; the air smells musty. It's such a squalid place. What's that over there? Is that rat droppings? There is no proper light in here; not enough light, and not enough air either. Carefully I continue walking through my grandfather's house, crossing the dark fishbone parquet into the former trophy room. Amon Goeth once had a sign put up here that said HE WHO SHOOTS FIRST LIVES LONGER.

I had wanted to see the house where my grandparents lived. A Polish tour guide whose address I found on the Internet told me that it still stood. A pensioner lives there now, and every now and then he shows individual visitors around. The tour guide called the man and arranged for me to see the house.

In the Płaszów neighborhood of Krakow, the only dilapidated house on quiet Heltmana Street stands out like a sore thumb against the other neat and tidy single-family homes. Some of its windowpanes are broken; the curtains are dirty; the house looks unlived-in. A large sign on the front of the house says SPRZEDAM. FOR SALE.

The front door still looks beautiful; the wood is decorated with ornaments, and the dark red paint has faded only a little. An unkempt man opens the door and leads me up a narrow stairway into the house. My tour guide Malgorzata Kieres—she's asked me to call her by her first name—translates his Polish for me. I haven't told Malgorzata why I am interested in the house; she thinks I am a tourist with a general interest in history.

I look around. The plaster is coming off the walls. There is hardly any furniture. But there is a coldness that creeps into your bones. And a stench. The ceilings are underpinned with wooden beams. I hope the house won't collapse on top of me and bury me beneath it.

Crumbling walls, holding up the past.

Over a year has gone by since I first found the book about my mother in the library. Since then I have read everything I could find about my grandfather and the Nazi era. I am haunted by the thought of him, I think about him constantly. Do I see him as a grandfather or as a historical character? He is both to me: Płaszów commandant Amon Goeth, and my grandfather.

When I was young I was very interested in the Holocaust. I went on a school trip from Munich to the Dachau concentration camp, and I devoured one book about the Nazi era after another, such as *When Hitler Stole Pink Rabbit*, *A Square of Sky*, and Anne Frank's *Diary of a Young Girl*. I saw the world through Anne Frank's eyes; I felt her fear but also her optimism and her hope.

The history teachers at my high school showed us documentaries about the liberation of the concentration camps, and we saw people who had been reduced to mere skeletons. I read book after book, looking for answers, to find out what drove the perpetrators to act the way they did, but in the end I gave up: Yes, I found some explanations, but I would never understand it completely. Finally, finished with the subject, I concluded that I would have behaved differently. I was different; today's Germans were different.

When I first arrived in Israel in my early twenties, I picked up books about Nazism again. Yet even there, where I was meeting the victims and their children and grandchildren on a daily basis, more important issues soon took over. I had read so much and asked so many people about it—I felt like I knew everything there was to know about the Holocaust. I was much more interested in the here and now: the Palestinian conflict, the threat of war.

I had thought I knew it all, but now, at nearly 40, I have to start all over.

One of the first books I pick up is a classic from 1967, *The Inability to Mourn* by Alexander and Margarete Mitscherlich. I like their approach; they look deep inside each person and try to understand without judgment. In their role as psychoanalysts, they regularly dealt with patients who were active members of the SS or other Nazi organizations before 1945. These people did not appear to have any sense of remorse or shame; they and their fellow Germans continued to live their lives as if the Third Reich had never existed. Reading the book with the knowledge of my family history, I think of my grandmother, who denied Amon Goeth's actions until the end.

The conclusion the Mitscherlichs drew at the end of the 1960s was that the Germans had denied their past and suppressed their guilt; ideally, the whole nation should have been in therapy. That conclusion no longer applies to today's Germans.

I also read books by other Nazi descendants, for example by Richard von Schirach, son of Reich Youth Leader Baldur von Schirach, and by Katrin Himmler, great-niece of Heinrich Himmler,

Reich Leader of the SS. Their family histories are of great interest to me, and I look for similarities.

I begin to dig deeper, I question family and friends. My adoptive mother's stepfather in Vienna, for example, served in Africa under Erwin Rommel. On long mountain walks he would tell us children anecdotes from that time, thrilling adventure stories of valiant warriors fighting in the desert, stories of how they collected the early morning dew from the tent sheets for drinking water, or how they once had to dig their car out of the sand dunes. At first we thought that our "Opa Vienna," as we used to call him, was Rommel's personal driver, but he put us right: He was only one of the drivers in the German Africa Corps. One day "the Limey got him," and he would tell us stories in his Viennese dialect about his time as a prisoner of war.

He only told us one horror story from the war: A soldier had been murdered—beheaded—and afterward his decapitated body was still running around like a headless chicken. That story always gave us the creeps.

When it came to talking about his superior, Opa only had words of praise. Rommel, the sly Desert Fox, was a "decent" Nazi? An urban legend. What skeletons are my adoptive family hiding in the closet?

Memories of discussions with my adoptive father are coming back to me. He was a liberal, often volunteered his services to friends and neighbors, and played an active part in the peace movement. On the subject of the Holocaust, however, he could not let go of the question of whether the number of murdered Jews was really accurate, or if it hadn't been less. He and his friends would argue fiercely about it. My adoptive brothers and I found the discussion unnecessary and didn't understand why this issue was so important to our father.

Suddenly I am not so sure anymore: Am I really so different? Have we really left everything behind us? What does it mean for me, for our time, that my grandfather was a war criminal?

My perception of time is changing. Events that happened a very long time ago are suddenly feeling very recent again. In the last few months I have read so much, have watched so many films; everything seems so immediate. Maybe it's because, to me, this old story is now very new, very fresh. Often, when I delve into this world my grandfather inhabited, it feels as if these crimes happened only yesterday.

And now I am standing here in this dilapidated villa in Krakow. I am not quite sure what I'm doing here, in this house, in this city. Does being here make any sense at all? I just know that I had needed to come to Krakow now. Shortly before I came I was in the hospital—I'd had a miscarriage.

I am feeling sad and exhausted. My therapist advised me not to travel to Krakow in my condition, but I had really wanted to make this trip. First I flew to Warsaw and then I took the train on to Krakow, the city where my grandfather was infamous, where it rained ashes at the end of the war when he had the remains of thousands of people cremated.

I want to see where my grandfather committed his murders. I want to get close to him—and then put some distance between him and me.

On the ground floor, the old man is now showing me the living room. This is where the parties were held, he says with a sweep of his arm. Here they sat, my grandfather and the other Nazis, drinking schnapps and wine. Oskar Schindler was there, too. The old man leads me onto the patio. He explains that my grandfather had some building work done, had balconies and patios added. The view of the countryside was important to him, he says.

Amon Goeth's former commandant's villa in Płaszów in 1995

The house must have been beautiful once; I like the style. Did my grandfather redesign the building himself? Was he interested in architecture like me? Why am I even thinking about whether we share the same tastes? Amon Goeth is not the kind of grandfather you want to find similarities with. The crimes he committed override everything else. In the book about my mother, I read that my grandmother used to gush about Amon Goeth's table manners, even long after the war was over. He was a real gentleman, she said.

* * *

A concentration camp commandant who set great store by table manners.

Emilie Schindler, Oskar Schindler's wife, later said about Amon Goeth that he had a "split personality." "On the one hand he played the gentleman, like every man from Vienna; on the other he subjected the Jews under his control to unrelenting terror. . . . He could kill people in cold blood and yet notice any false note on the classical records he played endlessly."

Amon Leopold Goeth was born in Vienna on December 11, 1908, the only child of a Catholic family of publishers. His parents Bertha and Amon Franz Goeth named him after his father and grandfather: Amon. In ancient Egypt, Amon was the ram-headed god of fertility. In Hebrew, Amon means *son of my people*. In the Old Testament, Amon was a king of Judaea who worshipped pagan gods and was killed by his servants.

Amon Goeth's parents came from a humble background but had come into money with their bookselling business. They could afford to live in a middle-class neighborhood, have a maid and eventually own a car, too. The Goeths sold religious literature, icons, and picture postcards. Later on they expanded into publishing, producing books about military history which mourned the Germans lost to World War I. Amon Goeth's father was often away traveling for the business while his mother managed the shop, and as a young boy Amon was often looked after by his childless aunt.

Amon, or "Mony" as he was often called, went to a private Catholic elementary school. He wasn't a very good student. His parents eventually sent him to a strict Catholic boarding school in the country. His biographer, historian Johannes Sachslehner, suggests that Goeth's future "tendency to play strange sadistic jokes" might stem from experiences he had during this time, but there is no evidence to support this claim.

Amon Goeth left the boarding school at the end of tenth grade against his parents' wishes. At 17 he was already enthralled by radical right wing ideas and had joined fascist youth organizations. He was athletic and reckless—characteristics that impressed his new friends.

In 1931 he became a member of the NSDAP, the National Socialist German Workers' Party, or Nazi party for short, and soon after he joined their security force, the SS.

Heinrich Himmler's SS, also responsible for experiments on humans and mass murder in the concentration camps, was regarded as the elite unit: "The best of the best, you couldn't be more Nazi if you tried," journalist Stephen Lebert once described the spirit of this corps in a nutshell. Hans Egon Holthusen wrote, in his 1966 confessional autobiography *Volunteering for the SS*, "This organization with their black uniform and death-head emblem was seen as elite, chic and elegant, which is why it was the organization of choice for the privileged youths who considered themselves too posh to go running around in the 'shitty-brown colored' outfit of the SA, the storm battalion."

The young Amon Goeth, unsuccessful at school and constantly pressured by his parents, was among those drawn to the idea of belonging to an elite. Later he would tell his live-in lover Ruth Irene Kalder that his parents had neglected him as a child and that he had turned his back on the middle-class values that they had tried to instill in him. It is true he returned to the family business for a short period of time, successfully publishing military history with his father. He even married a woman his parents introduced him to, although he wasn't in love. This "arranged marriage" soon ended in divorce.

An SS man has to start a family though, so Goeth married for a second time, this time to Anna Geiger, a sporty girl from Tyrol whom he had met at a motorcycle race. Since the aim of the marriage was above all the conception of healthy, "Aryan" offspring, the pair had to undergo a number of tests for the SS. For example, they had to have their pictures taken wearing only swimsuits to demonstrate their physical flawlessness. They were married by an SS man. Anna soon gave birth to a son, but the baby died after just a few months.

Shortly afterward, in March 1940, Amon Goeth reported for duty with the Waffen-SS, the military arm of the SS, and left Vienna for Poland. He was ambitious and climbed the ranks quickly. At first he was only charged with administrative tasks. An appraisal from 1941 states that he was an "SS man willing to make sacrifices, fit for service," "SS leadership material," and that the "overall racial image" was there, too. In 1942, in the Polish city of Lublin, Amon Goeth was given orders to establish labor camps to accommodate Jewish forced laborers.

In 1943, Heinrich Himmler delivered his infamous speech in front of top SS officials where he propagandized an ideology of hatred and contempt: "Whether other nations are living in prosperity or are starving to death interests me only insofar as we need them as slaves for our culture. . . . Whether or not 10,000 Russian women die of exhaustion during the construction of an anti-tank ditch interests me only insofar as the ditch is being finished. I also want to mention . . . a very difficult subject here. . . . I am talking about the evacuation of the Jews, the extermination of the Jewish people. . . . Most of you will know what it is like when 100 bodies are piled up together, when there are 500 or when there are 1,000. To have seen this through and to have stayed decent—with the exception of human weaknesses—that is what has made us tough."

Amon Goeth soon went on to prove his toughness. The SS taught him how to kill.

■   ■   ■

UPSTAIRS, THE OLD MAN UNLOCKS the door to the former bedroom. There are hooks in the ceiling. *This is where Amon Goeth did his exercises,* the old man claims. *Or maybe,* he adds with a wink, *he had a love-swing hanging from there.*

I step onto the balcony and look out over the hills covered in brushwood. A cold wind blows in my face. It is a rainy October day. The camp, surrounded by barbed wire and guarded by watchtowers, was located near the house. My grandfather could keep an eye on his prisoners; in the mornings it was only a short walk to work. That blurred photograph of Amon Goeth on the cover of the book about my mother—his open mouth, the bare chest, the rifle in his hand, wearing only shorts on his balcony—who took that photo? Was it my grandmother? Amon Goeth is said to have been proud of his firearms; he liked to carry them around with him. Did that impress my grandmother, or did it frighten her? What did she know? What did she suppress? I cannot imagine her living in this house, yet not being aware of what was happening in the camp. Amon Goeth is said to have beaten his maids. My grandmother must have seen or at least heard that, too. The house isn't that big.

After my arrival in Krakow the previous night, on my way to the hotel, I drove past Wawel Castle, the former residence of the kings of Poland, high above the Vistula. The castle was brightly lit. After the German invasion, Hans Frank, Hitler's governor of Poland, made himself at home there, living a life of luxury surrounded by servants, employing composers and chess players. I can imagine the life he had up there, how powerful he must have felt residing in that grand castle with its view over Krakow.

By comparison Amon Goeth's house looks very normal, almost modest. I had imagined it to be bigger, more ostentatious. I find it difficult to imagine that glamorous receptions were held here and that its owner was a man who was master of life and death for thousands of people. A man who thrived on having absolute power, and who wielded and relished this power absolutely.

*   *   *

Amon Goeth on the balcony of his villa

"I am your God," said Amon Goeth to the prisoners in his inaugural address as commandant of the Płaszów camp. "I dispatched 60,000 Jews in the district of Lublin. Now it's your turn."

In the Polish city of Lublin, Amon Goeth had worked for Odilo Globocnik, an SS man known for his brutality, whom Heinrich Himmler had charged with killing the Jews in occupied Poland. In December 1940, Globocnik updated Hans Frank on the goal he had set out with these words: "In this one year, I have obviously not been able to eradicate all the lice and all the Jews. But I am convinced that in the course of time it can be achieved. . . ."

When the "Final Solution" was being strategically coordinated at the Wannsee Conference on January 20, 1942, the deportation and mass murder of Polish Jews was already in full swing.

Goeth's superior Odilo Globocnik was co-responsible for the construction of concentration camps and the installation of gas chambers. In consultation with Adolf Eichmann he planned the factory-style murder of millions of people. In Poland, extermination camps were being commissioned: Belzec, Sobibor, Treblinka.

Soon Odilo Globocnik charged Amon Goeth with the liquidation of the ghettos. Liquidation meant rounding up the able ghetto population into forced labor; those too weak or too ill to work were shot, including children and the elderly. Historian Johannes Sachslehner describes the process as "blood-thirsty manhunts following a proven formula. . . . In the thick of it is Amon Goeth, who is soon entrusted with leading roles."

If he hadn't done so already, Goeth surely now discovered the lucrative side of genocide: Jews who offered him valuables such as furs, fine china, or jewelry were not killed immediately but were "allowed" to go to the labor camps.

Around this time Amon Goeth also started to drink more and more heavily.

Soon the ambitious Goeth was given more tasks: He was to lead the liquidation of the Krakow ghetto and establish a forced labor camp in Płaszów. In letters to his friends and to his father in Vienna he said, "Now I am the commandant at last."

On March 13 and 14, 1943, he ordered the clearance of the Krakow ghetto. Around 2,000 people are killed during these two days; a further 4,000 are deported, many to Auschwitz.

The survivors were taken to Amon Goeth's realm: Płaszów. Almost 200 acres in size, the camp was first a labor camp and later a concentration camp. The German occupiers had built it on Jewish cemeteries. They built barracks on top of the demolished graves and used the gravestones to pave the streets in the camp.

■ ■ ■

THE OLD MAN LEADS ME into the basement. "This is where the commandant stored his wine," he says. And then he points proudly to a rusty tub: "Amon Goeth's authentic bathtub."

Opposite the wine cellar and next to the kitchen was the maids' room. So this was Helen's place, here in the basement—Helen Rosenzweig, Amon Goeth's former Jewish maid from the American documentary I watched on TV the day after I discovered the book.

My mother met Helen here in this house. Ultimately it was a very sad encounter: Helen was shocked because my mother had such a striking resemblance to Amon Goeth. And even though Helen and my mother both try very hard, they cannot form a relationship with each other; history stands between them. Helen sees Amon Goeth in my mother.

In the film, when my mother tries to find an explanation for Amon Goeth's actions, Helen snaps angrily: "He was a monster. He was smiling and whistling when he came back from killing. He had the urge to kill, like an animal. It was obvious."

My brother Matthias has given me the documentary on DVD so that I can watch it again and again. At first I focused only on my mother and didn't pay much attention to Helen. The film begins with my mother writing a letter to Helen asking her for a meeting. In the letter she says that she imagines Helen might be afraid of meeting her—she herself is scared to meet Helen.

At the start, I wasn't so concerned about the actual contents of the letter. All I could think was, why does my mother spend so much time writing a letter to Helen? Why doesn't she write to me? Why does she share Helen's pain but not that of her own child?

Then gradually my feelings faded into the background and suddenly I saw Helen. I saw her, after all those years, returning anxiously to this house that used to be her dreadful prison. I saw how she is still plagued by her memories. She recounts how Amon Goeth used to beat the maids, how he pushed them down these very stairs, how he screamed at them and called them *slut, bitch, dirty Jewess.*

Helen's boyfriend was a member of the Jewish Resistance in the camp and was shot by Goeth. Helen also talks about the man she loved after the war, a camp survivor like herself. They were married for 35 years, moved to Florida and had children. Yet her husband could not get over the experience of the camp, and one day he took his own life. In his suicide note he wrote, "The memories haunt me every day. I just can't go on."

I am standing in the basement of my grandfather's house, in the darkness of Helen's room, where the only light comes from a small window. You can see a small patch of the garden. It was warm here; she didn't have to sleep on straw in the drafty barracks and was certain to have had more to eat than the other detainees. She didn't have to perform hard labor in the quarries like most of the other women in the camp; she wore a black dress with a white apron and served roast meat and wine. Yet she was living beneath the same roof as the man who could kill her at any time. She expected to die in this house.

■　■　■

When you saw Goeth, you saw death, one survivor said. The Płaszów camp became the stage for Amon Goeth's cruelty.

There are many eyewitness reports relating to this. Goeth's Jewish assistant, Mietek Pemper, described how once, in the middle of a dictation, the commandant suddenly grabbed his rifle, opened a window, and started shooting at prisoners. Pemper heard screams; Goeth then returned to his desk and asked calmly, as if nothing had happened, *Where were we?*

When Amon Goeth killed somebody, he would have their relatives killed, too, because he didn't want to see any "unhappy" faces in his camp.

In her memoirs, Płaszów survivor Stella Müller-Madej writes about Goeth, "If there was somebody he didn't like the look of, he'd grab him by the hair and shoot him on the spot. He was a giant of a man, a powerful, imposing figure with beautiful, gentle features and an even gentler expression on his face. So this is what a cruel, murdering monster looks like! How can that be?"

Goeth used publicly celebrated executions to crush any thought of escape or resistance the detainees might have had. Public hangings and shootings on the parade ground were accompanied by popular music. Larger groups of people were usually shot on a hill a little further away; the pit for the bodies was just below.

The Płaszów camp was growing, and the prisoners were now coming from further afield. Survivors from other ghettos, Polish prisoners, Romany from other camps, as well as Hungarian Jews all joined the Jews from the Krakow ghetto. Sometimes the Płaszów concentration camp held more than 20,000 detainees in its 180 barracks surrounded by two and half miles of barbed wire.

Within the SS, Amon Goeth was promoted to the rank of *Hauptsturmführer*, captain, which was an extraordinarily fast rise. He got rich on the prisoners' possessions and lived a life of luxury. He had a Jewish cobbler make new shoes for him every week and a pastry chef bake fancy cakes for him until he was piling on the pounds. In his villa, he held parties; alcohol, music, and women were offered in abundance to humor the SS men. Goeth owned riding horses and a number of cars; he enjoyed riding through the camp on his white horse or racing around the lanes in his BMW.

Mietek Pemper, the commandant's assistant, also took dictation of

Goeth's personal letters to his family in Vienna. Omitting the details of everyday life in the camp, he would ask his father about the progress of the business and his wife about the children: Anna Goeth had given birth to two more children, Ingeborg and Werner. When Amon Goeth learned that Werner was hitting his sister Ingeborg, he told Pemper to write in his letter to his wife Anna: "He must have got the hitting from me."

Eyewitnesses reported that Goeth would wear different accessories depending on his mood on a given day. If he put on white gloves or a white scarf, combined with a peaked cap or a Tyrolean hat, the detainees had to expect the worst. His two dogs, a Great Dane and an Alsatian mix, were called Rolf and Ralf, and he trained them to attack people on his command.

In 1944, Amon Goeth had children from the Płaszów camp herded onto trucks—to be transported to the gas chambers at Auschwitz. He ordered waltz music to be played over the loudspeakers in order to drown out the desperate cries of their parents.

In other words: Amon Goeth was perfect Hollywood material. Just as Adolf Eichmann was for many years the epitome of the callous, bureaucratic mastermind behind the scenes, denying any responsibility, Amon Goeth serves as the grotesquely excessive personification of the sadistic murderer. The image of the trigger-happy concentration camp commandant, accompanied by his two dogs trained to tear humans apart—it seems like a grim archetype, like a template for Paul Celan's poem *Death Fugue*. Steven Spielberg portrays Amon Goeth as a twisted psychopath, cruel, and at the same time almost laughable.

Other, big- and small-screen documentaries about Amon Goeth are often accompanied by ominous background music, but actually his crimes don't call for any embellishment.

Amon Goeth's crimes were so reprehensible that it seems easy enough to distance oneself from them. In his dissertation about concentration camp commandants, the Israeli historian and journalist Tom Segev writes:

Goeth would often ride around the camp on his white horse

"They were by no means Germans like all the other Germans, not even Nazis like all the other Nazis. They are not characterized by the banality of evil, but much more so by their inner identification with this evil. Most of the concentration camp commandants joined the Nazi movement in its early days . . . ; they had been vehement supporters of these radical rightwing politics right from its beginnings. The majority of Germans never even joined the Nazi party."

But Segev's theory might be too simple after all: The late Marcel Reich-Ranicki, literary critic and Holocaust survivor, had good reason to oppose the fact that famous Nazis such as Adolf Hitler are usually portrayed purely as monsters. "Of course Hitler was a human being," Reich-Ranicki said, adding, "What else would he have been? An elephant?"

It is very easy to demonize the prominent Nazis, to treat them like animals in a zoo: Look, aren't they cruel and perverted? It offers a way out of having to deal with one's own actions, one's family's actions—or indeed those of the many people who joined in on a small scale, those who no longer greeted their Jewish neighbors, and those who looked away and walked past when Jews were being beaten up in the streets and their businesses were destroyed.

■ ■ ■

THEY CALLED GOETH THE "BUTCHER OF PŁASZÓW." I keep on asking myself how it was that he became that way. I don't think that it was his childhood or even his hatred of the Jews. I think it was much more banal than that: In this world of men, killing was a contest, a kind of sport. It reached the point where killing a human being meant nothing more than swatting a fly. In the end the mind goes completely numb; death has entertainment value.

I have a terrible image in my head, which used to haunt me even in my sleep: It is said that Amon Goeth once caught a Jewish woman who was boiling potatoes in a large trough for the pigs—just as she, driven by hunger, ate one of the potatoes herself. He shot her in the head and ordered two men to throw the dying woman into the boiling water with the potatoes. One of them refused, so Goeth shot him, too. I don't know if this story is true or not, but I cannot get the image of this half-dead woman thrashing around in the boiling water out of my head.

These stories of how Amon Goeth considered himself superior, how he played music to accompany executions, used scarves and hats as props for his killings, how he played the master in his pathetic little villa—it would be comical if it wasn't so sad. He was a narcissist—but not just in the sense that he was in love with himself. He was a narcissist who felt on top of the world when he humiliated and degraded others.

I read that my grandmother used to idolize him: handsome Amon Goeth, the man of her dreams.

This is juxtaposed to the image of him that contemporary witnesses have painted: quick-tempered, cruel, irascible. His dogs. His exaggerated masculinity: commanding, dominating. Uniform, discipline, Fatherland.

My mother always saw the father in him, too, not just the concentration camp commandant. She is much closer to him than I am, even though she never met him. She was still a baby when he was hanged. Survivors of the camp have told her again and again how much she looks like him. How dreadful that must have been for her.

Do I look like him? My skin color is like a barrier between us. I imagine myself standing next to him. We are both tall: I am six foot, he was six foot four—a giant in those days.

He in his black uniform with its death-heads, me the black grandchild. What would he have said to a dark-skinned granddaughter, who speaks Hebrew on top of that? I would have been a disgrace, a bastard who brought dishonor to the family. I am sure my grandfather would have shot me.

My grandmother was never bothered by my skin color. She always seemed delighted to see me when I came to visit. No matter how little I was at the time: Children can sense if someone likes them, and she liked me. I've always felt so close to her. Yet she also held Amon Goeth when he came back from his killings. How could she share her bed and her home with him? She said she loved him, but is that a good enough excuse? Is it good enough for me? Was there anything loveable about Amon Goeth—is that even a permissible question?

When I look in the mirror I see two faces, mine and his. And a third, my mother's.

The three of us have the same determined chin, the same lines between the nose and the mouth.

Height, lines—those things are only external. But what about on the inside? How much of Amon Goeth do I have in me? How much of Amon Goeth does each of us have in us?

I think we all have a bit of him in us. To believe that I have more than others would be to think like a Nazi—to believe in the power of blood.

The quiet in the villa is suddenly broken: Malgorzata, the Polish woman who interprets the old man's Polish for me, reveals out of the blue that she once met Amon Goeth's daughter Monika. I ask her to tell me more and she says that my mother once came to visit the villa with a group of Polish schoolchildren. They were accompanied by a descendant of another Nazi, Niklas Frank, son of Hans Frank, governor-general of Hitler's occupied Poland.

Since Malgorzata doesn't know who I am I ask her what she thought of my mother. "I thought she was a bit strange, and sad," she replies. "Niklas Frank and Monika Goeth, neither of them could laugh." And then she tells us that here in this house Monika Goeth had touched a doorpost and said that she loved her father.

My mother's hand on the door. There are hundreds of German-speaking tour guides in Krakow, and I chose the one who has met my mother.

I tell Malgorzata who I am. At first she doesn't believe me, then she becomes bewildered and confused. I apologize to her. In order to find out more about my mother I had asked questions without revealing my identity. I say that I hope she understands my situation.

I had been determined to contact my mother before the end of the year. Now the year is almost over—it is well into fall, but I don't want to write to my mother until I feel better prepared for it.

In the documentary about her meeting with Amon Goeth's former maid Helen, my mother frequently cries. I can see the strain she's under because of her father's past. Krakow has a special meaning for her. I thought that I'd be able to understand my mother better if I, too, got to know this place.

The old man shows Malgorzata and me out. I pull the door firmly shut behind me.

There's another tour I've booked in Krakow: a *Schindler's List* tour.

I take a taxi to the meeting place in Kazimierz, the former Jewish quarter of Krakow. In the summer, Kazimierz is meant to be picturesque and charming; but today it seems dark and gloomy. The cobblestones are wet with rain. Our group of tourists visits the old Jewish cemetery, a synagogue, and a few locations from *Schindler's List*. We see idyllic courtyards and narrow alleys.

Many restaurants in Kazimierz serve gefilte fish and kosher meat. Pretty little cafés play traditional klezmer music—the rhythm of a long-lost time—around the clock. There is a morbid, museum-like feel to the whole district.

The narrow little alleys and the rough paving stones remind me of Mea Shearim, the Orthodox Jewish quarter of Jerusalem. The difference is, Jews actually live in Mea Shearim. Our tour guide tells us that before World War II there were 70,000 Jews in Krakow. Now there are only a few hundred. Most of the Jewish people walking through Kazimierz these days are visitors. There are six tourists in my group; I want to know where they are from. The response: Poland, USA, France. They want to know where I come from. "Germany," I say. "Ah!" they reply. I am glad we are not wearing name badges.

I still haven't really told anybody about my family history, apart from my husband, my adoptive family, and a close friend. Not because I think I need to be ashamed of it, but because I don't know how to deal with it. I find it hard to share my discovery. How could I put it? "Oh, by the way, I'm the granddaughter of a mass murderer"? I can't cope with my background myself, and I don't want to burden anybody else with it either. Not yet anyway.

Our small group moves on; we go over a little bridge across the Vistula river to the neighboring district of Podgorze. This is where the entire Jewish population of Krakow was crammed into a ghetto. The trams still ran through the middle of it, taking the people of Krakow across their city. Inside the ghetto, nobody was allowed to get in or out of the tram; there were no stops, and the doors and windows were locked for the duration. I wonder how the people of Krakow felt when they traveled through here.

Today a large office building stands in the square that used to be the center of the ghetto, and there is a bus depot, too. On the edge, a few sections of the ghetto walls remain. To add insult to injury, these tall walls surrounding the people behind them had arches at the top—they were built in the shape of Jewish tombstones. The message to the Jews was clear: You won't get out alive.

The victims are remembered at Ghetto Heroes Square. Empty, larger-than-life chairs dotted around the square are meant to convey a sense of the devastation in the ghetto after it had been liquidated: Everything laid to waste, no people in the streets, only furniture and other personal possessions that the Jews had been forced to leave behind. I find the installation too cold, too conceptual. Hundreds of people were killed during the clearing of the ghetto. Every chair represents 1,000 Jews who were murdered. The brutalities that were committed here remain abstract. But how can they be displayed? *Schindler's List* is very graphic, but survivors say that even Spielberg's film barely hints at the real horror that emanated from Amon Goeth.

※　※　※

Tadeusz Pankiewicz, a Polish druggist in the Krakow ghetto, described Goeth as a tall, handsome man with blue eyes, dressed in a black leather coat and carrying a riding crop. Survivors reported that during the liquidation of the ghetto Goeth tore little children from their mothers' arms and flung them to the ground.

Before the liquidation, 20,000 people were living in the ghetto in cramped conditions and in constant fear of death.

When Amon Goeth had the ghetto cleared on March 13 and 14, 1943, it had already been divided into two separate areas: Ghetto A was for

In front of the walls of the former Jewish ghetto in Krakow

those who had been declared fit for work and who were to be transported to the Płaszów camp; they were allowed to live for now. Ghetto B, separated from Ghetto A by barbed wire, was for the old, the young, and the ill; they were going to be killed.

Nobody was to get away. Goeth's people combed through all the alleys, looked in every apartment and under every bed. In the hospitals, the patients were shot in their beds. Tadeusz Pankiewicz described the scene of the aftermath like this: "It looks like a battlefield—thousands of abandoned bundles and suitcases . . . on pavements drenched in blood."

■　■　■

WE MOVE ON. It is raining, and we have to keep looking for shelter. A nice, elderly lady lets me share her umbrella. We walk beneath a drafty underpass and enter an industrial park. We stop in front of a gray, three-story office block dating from the 1930s: Oskar Schindler's former enamelware factory on Lipowa Street.

Today it has been turned into a museum dedicated to his memory. We visit the exhibition. It starts with photographs of Krakow in the early 1930s. There are women going for their daily walk and men on their way to synagogue. Then follows a portrayal of the German blitzkrieg against Poland and the immediate ostracism of the Jews. One picture shows German soldiers cutting off the sidelocks of an Orthodox Jew with a knife.

I am tired and worn out. I've been on my feet since the early morning. I would love to just sit down somewhere and get some rest, but the tour guide keeps on talking. I am losing my concentration and can no longer remember any of the details.

In the last room of the museum there is a recreation of the Płaszów camp. Little model barracks, even my grandfather's villa.

I take a close look: Again it is obvious how close Goeth's villa was to the camp and the prisoners' barracks. I find my grandmother's justifications less and less credible.

The exhibition covers Oskar Schindler, the man, only marginally. His story is told by means of photographs, documents, and his original furniture. In one room there is a large transparent cube filled with tin pots, bowls, and plates that were produced in his factory during the war. The installation is meant to symbolize the history of Schindler and his workers. Inside the cube are the names of the 1,200 Jewish slave laborers whose lives Oskar Schindler saved.

At the end of the exhibition there are two books, a black one and a white one. The white one is for the names of the people who helped save the Jews, the black one for the names of those who denounced and persecuted them. Two books representing two options: to save or to kill. Oskar Schindler or Amon Goeth. I don't like this simple division between good and evil.

Many Jews survived by going underground, thanks to help from relatives, friends, or work colleagues. These "quiet heroes" are not remembered often enough. Oskar Schindler was certainly not holier-than-thou but a rather ambiguous character. I find it hard to picture what he was really like.

■　■　■

Oskar Schindler and Amon Goeth: The two men were the same age and shared the same weakness for drink, parties, and women.

Both got rich on the back of the pogroms: Goeth by stealing from Jews and by killing them, by taking everything they had. Schindler by acquiring a factory in Krakow whose Jewish owners had been dispossessed and employing Jews from Goeth's camp as cheap labor.

Oskar Schindler, who had worked as an agent for the German counterintelligence in Poland, was a wartime profiteer at first; he came to Krakow to make a fortune. Later on he would spend most of the profits he had made on saving Jews.

The commandant Amon Goeth and the industrialist Oskar Schindler got on well together. Oskar Schindler needed cheap Jewish laborers and was therefore dependent on Amon Goeth's goodwill. Oskar Schindler called Goeth "Mony," brought him gifts, and introduced him to pretty women—among them Ruth Irene Kalder, who would eventually become Goeth's live-in lover.

Helen Rosenzweig, the Jewish maid in the Płaszów villa, reportedly said that Goeth believed that Schindler was his best friend. She, too, had been under that impression. True, Schindler had promised again and again to save her, "but then there he was again in his brown Nazi uniform taking part in Goeth's wild parties." There had been other factory owners who were helping Jewish workers and who depended on Goeth's goodwill. Yet they had not joined him in his revelries. "Schindler crossed limits that he didn't need to cross." Still, in the end she finds for Schindler: "Amon Goeth and Oskar Schindler, they both had power. One used it to kill, the other to save lives. Their example shows that everyone has a choice."

Steven Spielberg also plays with this motif in *Schindler's List*. He shows Amon Goeth as Schindler's evil twin. Both men appear to have been cut from the same wood, but their actions could not have been more different.

Goeth allowed Schindler to employ prisoners from the camp in his factory; he even allowed him to build an external camp for the workers at his factory where living conditions were much better than at Płaszów.

Amon Goeth's Jewish secretary Mietek Pemper and Oskar Schindler would sometimes hold secret meetings. Pemper said later that he had regarded Oskar Schindler as a savior from very early on. "No one other than Schindler was interested in our fate."

The Płaszów camp was run as a labor camp until the fall of 1943 when the SS administration decided to turn the last remaining labor camps into concentration camps. At the same time, more and more Polish camps which were not producing "strategic or pivotal" goods were being disbanded and the prisoners killed.

Consequently, Mietek Pemper conceived a plan: He wanted the Płaszów camp to be given the formal status of concentration camp because he was convinced that "the concentration camps will stay until the end of the war." Oskar Schindler claimed that apart from pots and pans his factory was also capable of making shell parts. Amon Goeth was just as keen on keeping "his" camp and presented his superiors with lists, manipulated by Mietek Pemper, of strategic goods produced at Płaszów. In fact, Płaszów officially operated as a concentration camp starting in January 1944. The prisoners were re-registered and were issued different clothes. New SS wardens arrived, and Amon Goeth was kept under tighter control. In his memoirs, Mietek Pemper explains that Goeth now needed written authorization from Berlin in order to torment his prisoners. He gives an example that highlights the bureaucracy of torture at the time: "A form had to be filled in with the requested number of lashes to the exposed buttocks." The camp's finances were also audited on a more regular basis now.

Amon Goeth had a look around other concentration camps and returned with new ideas, such as the tattooing of prisoners or the creation of a brothel for use by especially productive inmates. However, he never got a chance to implement these ideas.

By mid-1944 Płaszów was about to be disbanded. The Wehrmacht, the German army, was on the retreat, and the Red Army had entered Poland. In the summer of 1944, special SS task forces carried out a so-called "exhumation action" in Płaszów: All evidence of the exterminations was to be erased: The mass graves of the victims of the ghetto liquidation and other killings were dug up and the bodies burned. A putrid stench hung over the camp for weeks; the ashes were carried away by the truckload.

Emilie Schindler reported that in August 1944, her husband Oskar was worried about his laborers because Amon Goeth had decided to close Płaszów and send all the inmates to Auschwitz.

At the time, Oskar Schindler had his eye on an arms factory in Bruennlitz, near his hometown of Zwittau. His aim was to bring his workers there, to safety. Emilie Schindler described how her husband would bring Amon Goeth more and more expensive gifts. In the end, according to a number of different sources, the two men agreed upon a "deal": Amon Goeth would help Oskar Schindler smuggle "his" Jews to Bruennlitz, and Oskar Schindler would help Amon Goeth smuggle some of his valuable possessions to safety. Ultimately, however, Goeth's superiors in the SS also agreed to the transport of the "Schindler Jews" to Bruennlitz.

The list of people who were allowed to survive includes the names of around 800 men and 300 women. The circumstances behind each name's addition to the life-saving list are still unclear today. What is certain is that Marcel Goldberg, a Jewish camp inmate, took bribes and would swap names on the list for valuables.

After the war, Oskar Schindler did not meet with much success. Some of the Jews whose lives he had saved supported him financially. For his rescue of over 1,000 Jews, Schindler was named Righteous Among the Nations at a ceremony at the Israeli Holocaust memorial Yad Vashem in Jerusalem. He died in 1974 and was buried in Jerusalem.

There are many speculations as to what drove Schindler, what motives he had to save the Jews. Mietek Pemper sums Oskar Schindler up in these words: "This man, who had no special achievements to boast of, neither before the war nor after it, carried out a rescue operation, together with his wife, to which today, directly or indirectly, more than six

Oskar Schindler (2nd on the left) with some of his colleagues in Krakow, 1942

thousand people, including children and grandchildren, owe their lives. That is what's important. Nothing else matters."

■　■　■

OUR GUIDED TOUR OF THE MUSEUM has come to an end. I hang back to talk with the nice elderly lady from my group, the one who had offered me her umbrella. She is Jewish, from America, in her early seventies. She looks sporty, has short gray hair and bright eyes. I ask her if she has come to Krakow by herself. No, she replies, actually she's here with her husband. They are both Auschwitz survivors. Since they arrived in Poland, her husband has been gripped by a sudden fear, and he can't bring himself to revisit the scenes of his ordeal. He is distraught and stays in the hotel because he doesn't dare to go outdoors. That's why she went on her own— they had booked the tours before they came: Auschwitz yesterday, old Jewish Krakow today. She says it upsets her very much that her husband is suffering so much.

The story of the traumatized man who is too afraid to leave his hotel room moves me. I would like to cheer his wife up a little. I tell her that I have lived in Israel. She seems pleased and asks me what it was like for me there. We keep talking for a little while. She wants to know what I'm doing here, why I have come to Poland. Again I pretend that I am just a tourist with an interest in history. I offer her a lift in my taxi back toward Kazimierz, but she would rather walk a bit more.

For the second time today I have hidden my true identity. In the end I told Malgorzata, my tour guide, the whole story, but I couldn't tell it to this woman. I didn't want to tell her why I am here. There wouldn't have been enough time to explain everything. The knowledge would only have left her upset. She would have gone back to her husband in the hotel confused and maybe even troubled. But I am not comfortable with this secrecy either.

I will probably never see this friendly Jewish lady again, but sooner or later I will have to confide in my friends in Israel.

I head toward the Rynek, the magnificent medieval market square in the center of Krakow's Old Town. There is nothing gloomy about this place, there are no winding little alleys like in Kazimierz; here, everything is majestic and open. I browse the market stalls, looking for a bunch of flowers. I want something bright, but nothing gaudy. Mainly whites, with a mix of small and large flowers. I end up arranging my own.

■ ■ ■

During the German occupation, the Rynek in the center of Krakow was renamed Adolf Hitler Square. The Germans were already on the retreat

Amon Goeth (left) in Krakow on his way to court where he would be sentenced to death in September 1946

in Poland when the Płaszów commandant was arrested: The SS had discovered that Amon Goeth was secreting valuable assets away from Płaszów, so they initiated proceedings against their own man.

Goeth was accused of corruption and abuse of his position. He spent some time in Munich's Stadelheim prison but was soon released.

After a short stint at the front, Goeth was admitted to a hospital in the Bavarian town of Bad Toelz. His health had weakened; he was suffering from diabetes and had liver and kidney problems.

On April 30, 1945, the American army marched into Munich. On May 4, Amon Goeth—wearing only a Wehrmacht uniform and therefore not immediately identifiable as a member of the SS—was arrested in Bad Toelz. He gave a false name and claimed that he was a common soldier returning from the front. Meanwhile, his wife was filing for divorce in Vienna; she had heard of his affair with Ruth Irene Kalder.

The pregnant Kalder and her mother Agnes fled first to Vienna and then to Bad Toelz. On November 7, 1945, Goeth and Kalder's daughter, Monika, was born in Bad Toelz.

By then, Goeth had been taken to a detention camp on the grounds of the former Dachau concentration camp near Munich. From there, in January 1946, he wrote his last letter to Ruth Irene Kalder: "Dearest Ruth, Thank you for your letter and parcel. You poor thing, you have been through so much. . . . The food here is such that I only weigh 70kg now. It's enough. . . . Everything will be all right. Don't worry. . . . Lots of kisses for you and Monika and give my love to Omi. Love, Mony."

It wasn't long until the American investigators discovered who Goeth really was. Four former Płaszów prisoners identified him as the commandant of the camp. When they saw Amon Goeth in the company of American soldiers, one of the four witnesses greeted him with the words, "Herr Commandant, four Jewish pigs reporting for duty!"

Amon Goeth was extradited to Poland, together with the former commandant of the Auschwitz concentration camp, Rudolf Hoess. Goeth and Hoess arrived at Krakow's main station on July 30, 1946. An angry mob awaited them, but it wasn't Rudolf Hoess they were after, the man who had sent hundreds of thousands to their deaths in the gas chambers. It was Amon Goeth whom they wanted to lynch, the "Butcher of Płaszów."

Goeth was put on trial in Krakow in late August 1946. It was the first big trial of its kind in Poland, and it was going to last just a few days. The many spectators did not all fit into the courtroom, so the proceedings were broadcast to the outside via loudspeakers. Hundreds of listeners gathered on the lawns opposite the courthouse.

Goeth was charged with genocide. Among other crimes, he was accused of being responsible for the deaths of around 8,000 people at the Płaszów camp as well as the killings of 2,000 more during the liquidation of the Krakow ghetto. This was in addition to the murder of hundreds of people during the liquidation of the ghettos in Tarnow and Szebnie in the fall of 1943. Furthermore, he was charged with taking unlawful possession of the victims' valuables. Confronted with the number of witnesses for the prosecution, Amon Goeth reportedly called out: "What? So many Jews? And we were always told there'd be none left."

When Goeth was asked if he pleaded guilty, he retorted with a loud "No." At the trial he denied his crimes, giving the names of other SS men and saying that they had been responsible for the killings. He had only been following orders, he said, he had been a common soldier and had not given any orders himself. When witnesses described the murders in the camp, he looked away indifferently or tried to prove that they were making false claims. He called Oskar Schindler as a witness for the defense, but he did not turn up.

Absurdly, he also called his former secretary Mietek Pemper as a witness, but Pemper, who had witnessed many of Goeth's crimes, spoke against him rather than for him.

The Polish prosecutor demanded the death penalty and said in his summation to the jury: "You are being asked to judge a man who has become a legend during his lifetime for being the modern incarnation of the biblical Satan."

Amon Goeth was indeed sentenced to death. He appealed for clemency and asked for the death penalty to be changed into a prison sentence. He wanted to prove that he could be a useful member of society. The appeal was denied.

On September 13, 1946, Amon Goeth was led to the gallows. His last words were "Heil Hitler."

■　※　▓

THERE ARE MANY THINGS I'd like to ask my grandmother. I think it would have been worthwhile to press her for answers—she had some dents in her armor and might have been willing to talk.

As for my grandfather, I have few questions for him. Those images of his execution, of his arm raised defiantly in the air, the Hitler salute his parting gesture from life. If he had ever shown any signs of remorse, I would have liked to question him. Yet, as it was, I think it would have been pointless. He never admitted his guilt. At his trial he lied right up to the end.

I go to visit the former site of the Płaszów concentration camp. Today the hilly ground where the camp used to be is turfed over. There is nothing here to recall the barbed-wire fences, the watchtowers, the quarry where the prisoners toiled, the barracks, the mass graves. Only green grass, in between a McDonald's restaurant and a busy highway. In the distance, socialist prefab buildings loom against the sky.

High on a hill, visible from afar, stands the memorial: a larger-than-life sculpture of people with bowed heads, carved from light-colored stone. Where their hearts should be is a gaping hole.

I am surprised. I can still picture this setting from *Schindler's List*. Everything seemed so real, so alive. Now there are no moving pictures, only stones.

The camp is history; my grandfather has long been dead.

I hold onto my flowers and climb the wide steps up to the plateau where the memorial stands. From up here I get a much better view of the area. The site looks abandoned and neglected. Without the informational displays, one would never guess at the atrocities that were committed here all those years ago.

People are jogging by in the drizzling rain; in the distance I can make out others walking their dogs. They probably come here every day, grateful for this park's existence.

All alone I stand in front of the memorial. Few people come here at this time of year.

The memorial for the victims of the Płaszów camp

Reverently, I touch the cold stone with my hand, just like I did at the Wailing Wall in Jerusalem.

During the last few months I've been asking myself, *Who am I?* I'm not so sure anymore: Am I still Jennifer, or am I only Jennifer, the granddaughter of Amon Goeth, now? What counts in my life?

I can't just shove my grandfather's past into a box and put a lid on it. I can't just say, *That's the past, it's over, it doesn't affect me anymore.* That would be a betrayal of the victims.

I have come here as I would come to a grave. A grave is a place to care for and to return to in order to honor the dead.

When somebody dies it is not necessarily essential to go to their funeral. You can say good-bye privately. Yet the visit to the grave is a sign, an important ritual, which is why I have come here today. I want to pay my respects to the victims. To show that I will never forget them.

Slowly I lay down the flowers and sit on the grass. Only now do I realize that a group of people has gathered in front of the memorial. Children are running over the grass. A school class from Israel. I listen; it sounds dear and familiar.

CHAPTER 3

THE COMMANDANT'S MISTRESS:
MY GRANDMOTHER RUTH IRENE KALDER

It was a wonderful time. My Amon was king, I was his queen.
Who wouldn't have relished that?

—RUTH IRENE GOETH IN 1975,

about her time spent with concentration camp commandant Amon Goeth

HOW MUCH DID MY GRANDMOTHER KNOW?
Before my visit to the villa I had convinced myself that she probably wasn't aware of everything that was going on. Before going to Krakow I had imagined rambling grounds, a massive house. The shots fired in the camp would have been too far away, the screams of the maids being terrorized by my grandfather too quiet to hear.

Only it wasn't like that. My grandmother was right in the thick of it. The house was small, the camp not far away.

Was my grandmother not just blinded by love but deafened by it, too?

Where was my grandmother's compassion? People were dying a few hundred yards away, and there she was, reveling with Amon Goeth.

My grandfather has long been dead, but I knew my grandmother. When I was a small child, she was the person who mattered most to me. I had little, if anything, to hold on to. She liked me, and that meant a lot.

To me, she radiated kindness. Whenever I think of her, I feel safe and secure again.

And then I read the book about my mother, and now I am learning all these things about my grandmother that blatantly contradict the image I have of her.

If it hadn't been for her, maybe discovering Amon Goeth in my family tree wouldn't have been such a shock. I could have regarded him more as a historical figure; I might not have taken it quite so personally. Yes, he is my grandfather, but he never pushed my stroller or held my hand. But my grandmother did.

I feel so close to her, which is why I cannot simply shove the image of Amon Goeth to some distant place in history.

Literature written about the descendants of Nazi criminals sometimes differentiates between those who knew the relative in question and those who didn't. Some authors conclude that those who never met their Nazi ancestor are generally less troubled by their past. What they ignore, however, is that those born after the event often still know relatives who loved their Nazi ancestor. The living are their connection to the dead.

My mother was ten months old, a baby, when Amon Goeth was executed. Yet the book about her clearly shows that she is very distressed by the past. She has the same connection to him as I do: Ruth Irene—her mother, my grandmother. The woman who had a picture of Amon Goeth hanging above her bed until her death. My grandmother later said that he was the most important man in her life. What drew her to him?

The large bus I am on is full of people, but nobody is talking much. We drive along roads lined with low houses; the route is dotted with little villages. The pavement is wet. It has been raining again. I wish the sun would break through the clouds—the place where I'm heading is gloomy enough as it is.

It is my second and last day in Poland, and I am on my way to

Auschwitz. The former concentration camp is only an hour's drive away from Krakow. I have never been there before, even though Auschwitz is both a powerful symbol and harrowing relic of the Holocaust. Visiting these places is an altogether different experience from just reading about them. Auschwitz is where Amon Goeth sent thousands of Płaszów prisoners—straight to the gas chambers. Did he talk to my grandmother about it? Maybe not, but she must have known it nonetheless.

The more I try to understand who she really was, the harder I find it to stay objective.

■   ■   ■

Jennifer Teege's grandmother Ruth Irene Kalder, later Ruth Irene Goeth, was 25 years old when she met Amon Goeth. She came from Gleiwitz in Upper Silesia. Her father owned a driving school and was a member of the Nazi party. Ruth Irene Kalder had qualified as a beautician and had attended a drama school in Essen. During her time there, she allegedly had a relationship with an older man, got pregnant, and subsequently had an abortion.

In Krakow she was working as a secretary for the Wehrmacht. According to Goeth's biographer Johannes Sachslehner, she "had a reputation for not being disinclined to little adventures with men in uniform." She became friendly with the industrialist Oskar Schindler and carried out some secretarial jobs for him. One evening in the spring of 1943, Schindler took her along to a dinner with Amon Goeth.

Later, in interviews and when talking to her daughter Monika, Ruth Irene Kalder described meeting the concentration camp commandant as love at first sight. Amon Goeth was big and strong, she said, "a true dream for any secretary." She had "eyes for nobody but this man," adding that he was funny, intelligent, and well read, "the ideal man, like Clark Gable as Rhett Butler in *Gone with the Wind*."

In an interview with the Israeli historian and journalist Tom Segev in 1975, Ruth Irene Kalder also claimed that she was supposed to flirt with Amon Goeth in order to cement his good relationship with Oskar Schindler, who relied on Jewish laborers sourced from Goeth's camp. "My job as the pretty secretary was to win over his heart so that he would continue to provide us with these workers, since the Jews were now under the control of the camp commandant."

The petite, dark-haired young woman instantly hit it off with Amon Goeth. Ruth Irene Kalder recalled that they were soon on first-name terms; as she was leaving, Goeth said to her, "I'll call you." When he hadn't been in touch after a few days, she called him: "You said you'd call, I'm still waiting." Goeth was surprised but also suspicious. He was under the impression that she was Oskar Schindler's girlfriend and had called to spy on him. She reassured him that she was only friends with Schindler and arranged to meet him in Płaszów.

She soon became Goeth's permanent girlfriend; he gave her the pet name *Majola*. Her love for Amon Goeth brought her to the grounds of a concentration camp, into the house of its commandant.

Goeth's former Jewish maid Helen Rosenzweig has described Ruth Irene Kalder as a "beautiful young woman, with dark hair and wonderfully milky skin. She must have been very much in love with Goeth, for she was always gazing at him."

Kalder ignored her beloved Amon's less loveable side. According to Helen Rosenzweig, Ruth Irene Kalder did not want to know what was going on in the camp: "Most of the time she was busy mixing egg yolk with cucumber and yogurt, and then she would lie around with a cucumber mask on her face. She would turn the music way up so that she couldn't hear the shots."

Film director Steven Spielberg shows Ruth Irene Kalder burying her head in her pillow while Amon Goeth is shooting from his balcony.

Ruth Irene's mother, Agnes Kalder, visited her daughter at Płaszów once. Agnes was horrified to see the environment where her daughter was living and returned home early.

Ruth Irene Kalder, however, enjoyed her life of luxury at the commandant's side. She later told her daughter Monika that she and Goeth would often start the day with a horseback ride. Afterward, she would apply her extensive makeup. After breakfast she would give the orders for lunch: plenty of meat and alcohol for Amon Goeth, cake and fruit for dessert. In the afternoons, Ruth Irene Kalder would go for another ride, listen to

Ruth Irene Kalder, photographed by Amon Goeth, with Rolf, the Great Dane that Goeth trained to tear humans apart, and her own lapdog

music, or play tennis with the girlfriends and wives of the other SS men. In the evenings there would often be company. Amon Goeth and his girlfriend especially enjoyed having the Rosner brothers, Jewish musicians from the camp, perform for them. On those occasions, Hermann and Poldek Rosner would swap their prison clothes for elegant suits and play the violin and the accordion for Goeth and his guests. And Ruth Irene Kalder, dressed in fine clothes from Krakow shops, would play the lady of the house.

One photo from Płaszów shows Ruth Irene posing in an elegant riding dress in front of the somber barracks and barbed-wire fences as if she were modeling the latest fashion on the Champs-Élysées. In other photos she is sunbathing in a swimsuit on the patio of the commandant's villa. Another picture shows her in a stylish hat and coat, standing with her little black lapdog on one side and Goeth's favorite dog, his spotted Great Dane Rolf, on the other. Presumably, Amon Goeth took this picture of her.

■　■　■

FOR ALL THESE YEARS, I've had only one photograph of my grandmother. It shows her wearing a long, flowery dress, her hair combed into a beehive, the golden bangle on her arm twinkling in the sun. She is standing on the grass in the English Garden in Munich. A dachshund is playing behind her, a red ball lying in the grass. She is smiling at the camera and looks young, happy, and relaxed. It is a lovely, natural photo, which I have always treasured.

Now I am finding very different pictures of her, in the book about my mother and on the Internet. Looking at these pictures of her, posing with a dog that would attack people on Goeth's command—it is unbearable; it is too upsetting. I am exposing myself to a lot, but I can't and won't look at these pictures. How could she touch this dog, how could she bear having it near her? After all, it wasn't a pet, but an animal trained to kill.

I cannot reconcile these pictures with my image of her.

I do not grieve for my grandfather, but I do for my grandmother. I grieve for the person she never really was.

She was always good to me, which is why I always thought of her as a good person. As a child, you cannot imagine that the person you love could have another side, a darker one.

I really wish that my memories of her had not been tarnished. Why couldn't she have been just an ordinary grandmother—a nice lady who died one day?

I had always thought of Irene as one of my three grandmothers, the other two being my adoptive grandmothers, whom I called *Oma Vienna* and *Oma Bochum*.

Oma Bochum was my adoptive father's mother. She was very short, had gray curly hair—a typical grandmother's perm—and an energetic, scurrying walk. She would always wear skirts, covered with an apron to keep them clean. Whenever she left the house she would change into her orthopedic pumps with flat heels—the ones I called "click-clack shoes" as a child. When I went to visit her in Bochum with my adoptive family, we would accompany her to the market or to the butcher's, or help with her gardening. I was never too enthusiastic about planting vegetables or picking fruit, but I loved the results: The shelves in her basement were loaded with jars of stewed fruit. At dinnertime, a gong would summon us to the table.

She was very disciplined and also a little strict, not a cuddly grandmother. Yet she had a big heart. Even though my Oma Bochum had two children of her own—my adoptive father and his sister—she regarded it her Christian duty to include other children in her family. Helping orphaned or neglected children was a living family tradition. Growing up with a number of foster siblings was a matter of course for my adoptive father, which is why it later seemed natural to him to take in a foster child—me.

My Oma Bochum was an active member of the Protestant Church and very popular in the community. She would regularly visit the grave of her late husband, who had died young. She normally went to church on Sundays, and eventually she died there, too: She had a heart attack one Sunday in the middle of a service.

My Oma Vienna, my adoptive mother's mother, was also short, but very plump. She exuded something motherly, calming. She was always impeccably dressed and enjoyed wearing silk dresses and fur-trimmed coats. As a child, I often stayed with her; I preferred Vienna to Bochum—the city was more exciting. Oma Vienna would sometimes behave like a small child herself: Once we played a trick on Opa and pretended that we had all run away, Oma and the children. Opa played along, pretending to be really worried.

Only Christmastime was odd with Oma Vienna. We would all sing carols under the tree, but she wouldn't join in for fear of striking the wrong notes.

We often went on vacation with our grandparents from Vienna, too: skiing in the winter, hiking in the Austrian Alps, or camping by the sea in Italy in the summer. Opa sometimes told us wartime stories about his time with Rommel in Africa. Oma never talked to us about the war. In 1945, she had fled the area that is now the Czech Republic for Vienna. The journey was a terrible experience for her, but she would never discuss it.

And then there is my fourth grandmother, in Nigeria—my other biological grandmother besides Irene. I don't know much about her. I met my father once, when I was 28. He told me that, when my mother wanted to give me away to the orphanage, he had suggested that I might as well live with his mother in Nigeria. He would have preferred that to the orphanage, but my mother didn't like the idea. I guess she wasn't ready then to give me up entirely. While I was at the orphanage, my mother could still visit me, and she would still have had the option to take me back.

I imagine my African grandmother as a tall, proud woman, a strict matriarch. I find it remarkable that she would have been prepared to take me in. For that I am very grateful to her, and I sometimes wonder, what if . . . ?

I have never compared my grandmothers to each other, during my childhood or later. They were far too dissimilar for that. I had separate relationships with each of them, and each was important in her own way.

Nonetheless, Irene occupied a special place in my heart. She was one of the first people I was attached to as a child.

When I was seven and my adoption became official, my adoptive parents broke off all contact with my mother; they thought that it would be best for me. With that, my grandmother also disappeared from my life. She left behind a gap, I missed her.

I was 13 when last I heard about her: My adoptive parents told me that my grandmother had died. They had seen the obituary notice in the newspaper. It didn't mention that she had killed herself.

I didn't ask any questions. My biological family was not a subject we talked about in my new family. There was a deep and stony silence—a tacit agreement between my adoptive parents and me not to mention my mother or grandmother. Not that my adoptive parents could have told me much about them anyway; they didn't know anything.

I remember feeling sad when I heard about my grandmother's death. I had always hoped to see her again one day, but now she was gone for good.

Before I came across the book in the library, all I had were my memories: My grandmother enjoyed my company. With my mother I often felt that I wasn't welcome—my mother would pull me along by the arm when she was impatient, but Irene never did.

I remember only one exchange with my grandmother that confused me: For some reason I was feeling sad, but she was very unsympathetic and told me not to cry. I didn't understand what my grandmother had against tears.

She was not your classic grandmother; I wasn't even allowed to call her *Oma*, just Irene. Maybe she didn't want to be considered old. It's been said that she paid a lot of attention to her looks—her appearance was very important to her. Even my mother called her only by her first name; that's what it says in the book.

I remember her apartment in Schwind Strasse, in Schwabing. We would usually sit in her open-plan kitchen with the American radio station AFN turned on. I still enjoy listening to English-language radio stations; in Hamburg I often tuned into a British military station, and in Israel I'd listen to the Voice of Peace.

She didn't have a living room, as such. At home with my adoptive family in the Waldtrudering suburb of Munich, we would hang out on the sofa in the living room, and we would wear comfortable "house clothes" indoors. This would have been unthinkable at Irene's. It is true that I always felt at ease with my grandmother, but never quite at home: I was still a visitor. She was always elegantly dressed and nicely made up—everything was a little formal. The kitchen was always clean and tidy; I never saw her cooking or baking.

Unfortunately I have far too few concrete memories of her; I think of her as a child would: someone who cares, someone who protects.

Whenever my mother picked me up from the orphanage—or later, from my foster family—and dropped me off at my grandmother's, it meant that I didn't have to go my mother's place in Hasenbergl.

It wasn't like my mother had a happy family at home: Her then-husband was a drunkard and a wife-beater, and I felt constantly threatened by him. I never knew whether he was going to be there or not. If he was out, I'd hope that he wouldn't come back. I was always listening for the sound of his key in the lock or his footsteps in the corridor.

At my grandmother's, I felt safe. When I entered her kitchen, everything was all right.

■　■　■

Helen Rosenzweig, Goeth's former Jewish maid, tells this story about Ruth Irene Kalder: "Once she came down to see us in the kitchen. She reached out her hands to us and said: 'If I could send you home I would, but it's not in my power.'"

In Amon Goeth's villa, the maids, Helen Hirsch and Helen Rosenzweig, were subjected to constant abuse: He summoned them by shouting or by ringing a bell that could be heard all over the house. Often he would beat them if they didn't come running fast enough. One of those beatings left Helen Hirsch with a burst left eardrum; she remained deaf in that ear. Helen Rosenzweig has described how Goeth pushed her down the stairs countless times. "In his house, at his mercy, I lost all fear of death. It was like living under the gallows, twenty-four hours a day."

Ruth Irene Kalder later told her daughter Monika that she once intervened when Goeth was threatening to beat one of the maids with a bull pizzle—a dried bull's penis that was used as a flogging tool in the concentration camps. In the ensuing struggle, Amon ended up hitting Ruth, which he felt awful about. He came close to tears, she said, and apologized over and over, and after that he never again used a bull pizzle in the house. Ruth Irene also told her daughter another grotesque anecdote: She once threatened not to sleep with Goeth anymore "if he didn't stop shooting at the Jews." Apparently it worked.

Helen Rosenzweig felt she had spotted a "shred of humanity" in Ruth Irene Kalder. She remembers, for example, that Ruth would make it a point to praise the maids in front of Amon Goeth and that she always treated them with respect.

When Helen Rosenzweig's sisters were to be transported from Płaszów—presumably to Auschwitz—Helen Hirsch ran to Ruth Irene Kalder and begged her to prevent their deportation. At first, she refused: "Please don't ask me to do this!" Eventually, however, she caved in and called the camp police to stop the deportation of the Rosenzweig sisters. When Ruth Irene confessed her unauthorized rescue mission to Goeth, he was furious. According to Helen Hirsch, he came running to the kitchen with his rifle to find the maids, but eventually he calmed down.

Helen Hirsch also reported that the inebriated Goeth once tried to sexually assault her. Ruth Irene Kalder heard her cries and came running to her rescue. Goeth then let her go.

There are a number of eyewitnesses who remember that Ruth Irene Kalder tried to exert a moderating influence on Amon Goeth's behavior. She is said to have taken a stand for individual prisoners and to have prevented the torture and shooting of a number of inmates. In her presence, Amon Goeth is said to have been more restrained and mild-mannered. In another example, according to contemporary witnesses, she once called Goeth away from the parade ground while he was having prisoners whipped. Ruth Irene Kalder, however, would later claim that she never set foot in the camp.

Emilie Schindler recounted that around the middle of 1944, her husband Oskar reported that Goeth was getting tired of his girlfriend; the woman was "too peace-loving" and always trying "to dissuade him from his sadistic excesses."

That said, the fact that Ruth Irene Kalder sometimes half-heartedly tried to help the victims goes to show that she knew how Amon Goeth was treating them and what crimes were being committed in the camp.

In his autobiography, Goeth's secretary Mietek Pemper writes that Ruth Irene Kalder would sometimes type highly confidential documents for Goeth. Pemper believes that she also helped compile a list of prisoners to be executed.

Later, Ruth Irene Kalder would often stress two things, namely that Płaszów had only ever been a labor camp, not a death camp, and that there had been only adults in the camp, and no children.

Nonetheless, she told her daughter Monika that she had once observed children being transported from the camp by truck. Monika Goeth has said that her mother could not stop thinking about those children and that she believes her mother put her memories of the event to paper.

The trucks Ruth Irene Kalder remembered were probably those which took the children from Płaszów to Auschwitz on May 14, 1944. Goeth needed to create space in his camp for a number of Hungarian Jews who were due to arrive. Accordingly, he wrote in a letter to an SS leader, he had to "purge" the camp of its old, ill, and weak inmates, as well as its children, and thereby liquidate all the "unproductive elements." In other words, he was going to deport the weak and the ill from Płaszów to the gas chambers at the nearby camp of Auschwitz, for "special treatment."

Notices were put up at the parade ground proclaiming "Appropriate work for every prisoner." Loudspeakers blared out cheery tunes as the prisoners were told to undress and parade past the camp doctors. According to an eyewitness, among them was Josef Mengele, the infamous camp doctor from Auschwitz, who had come specially and was noting down the names of the children. A week later, the result of this so-called "health action" was announced. Those who were to be removed to Auschwitz had to gather on one side of the square: around 1,200 people in total, including about 250 children.

Płaszów survivor Stella Mueller-Madej describes the scene as the children were being herded onto the truck: "The whole place is pandemonium. Fathers and mothers are sobbing. The children, who until then had been silent like dolls and frozen with horror, are now screaming and pleading . . . They are crying for help . . . A very young child tries to crawl away to safety on all fours. A female guard . . . grabs . . . her by her hands and throws her little body onto the truck bed like a sack of potatoes. It is unbearable. Everyone on the parade ground is crying, the whips are lashing down, the dogs are barking . . . At that moment the loudspeakers start to play waltz music . . . and the trucks head off to the camp gate."

Shortly after their arrival at Auschwitz, the children were killed.

■   ■   ■

JUST AS MY GRANDMOTHER, HER WHOLE LIFE, made excuses for Amon Goeth and romanticized him, I tended to regard her too favorably in the beginning of my research. I told myself, "She didn't do anyone any harm. She was not actively involved in his deeds."

I knew so little about my grandmother. I saw my mother once again in my early twenties, but my grandmother had already passed away by then. When reading the book about my mother, I scrutinized the pictures of Irene closely—at first only the private ones from the years after the war, and then later the historic ones, too. Sometimes I can see myself in her.

I also love the good life. I drive a comfortable car, enjoy living in a big house, and appreciate modern conveniences. Like my grandmother, I like beautiful things, and I don't mind if they cost a little extra. But surely the question is: How high is the price?

I don't think it was just the status and the money that kept my grandmother in Płaszów—she undeniably enjoyed living in prosperity with Goeth, but I doubt that the luxury lifestyle alone would have been enough. After the war, she lived a much more modest life.

I think she was madly in love with Amon Goeth. Maybe she was also fascinated by his power, but there must have been more, some sort of inescapable pull or need, which blocked out everything else.

My grandmother never married or had a long-term relationship later in life. No matter who drifted in or out of her life after the war, Amon's photograph always hung in the same spot—another reason I think her relationship with Amon Goeth was based on more than a mere cost-benefit calculation.

I know how this feels, this evidently boundless love, because I'm exactly the same way. When I love someone, I love unconditionally. In this I can understand my grandmother. When I'm in love with a man, I give him carte blanche: In theory, he can do as he likes; he will always have a special place in my heart. I won't tell him so, of course, and it doesn't mean that I will always tolerate or approve of his behavior, but the love will always be there.

That raises the question, what would I have done in my grandmother's place? Could I have fallen for this sadist of a man? I can't give a straight answer to that, but just the thought of someone beating his servants with a bull pizzle is enough to turn my stomach.

By way of apology for my grandmother, my mother has said that the camp was not visible from the bedroom at the villa. The Jews in the camp allegedly said of my grandmother: "She's one of us." Her name was Ruth, after all, a Jewish name.

Am I to believe that? Or am I just glad to have an excuse? I am of two minds: On the one hand, I want to sustain the lovely image of my grandmother. On the other hand, I want to know the truth. At college I used to gather reference material and compare the different sources. What mattered in the end were not my assumptions, but the hard facts. It is the same with my grandmother: I have gathered a lot of material about her in order to gain a better understanding of her.

I am no judge; it is not my place to pass judgment on her. I just want to see her as who she really was.

My first reaction when I read about her trying to help the victims was one of relief. I thought, "She wasn't like my grandfather, maybe she was on the side of good." But now I am ashamed of this thought.

I try to picture the scene with the maids again: My grandmother standing in the kitchen and telling Helen, who had to fear for her life every moment of every day, that she would help her if only she could. There is a great coldness in that, too. She took a step toward Helen, but then abandoned her after all.

She had seen the maids' suffering; she was aware that she was caught in a predicament. But that's the fatal thing: She could tell right from wrong. She could have made a choice. But she was too selfish to let this inner conflict come to the surface.

She felt pity for others and even helped some of them. But is that enough? Absolutely not. I don't care whether she intervened a hundred times or a thousand. The fact is, it didn't change her priorities. In the end, she was just looking out for herself.

I believe that there is a difference between me and my grandmother, and a very fundamental one: I could never live with a murderer; I could never bear to be with a man who took pleasure in tormenting other people.

■ ■ ■

Whenever Jennifer Teege talks about her grandmother, her voice goes soft and her eyes beam.

Her feelings fluctuate between rejection and affection, attack and defense. She cannot get a handle on who her grandmother really was.

"I had no idea." Ruth Irene Kalder would repeat this sentence often after the war. It is a sentence that many young Germans have grown up with: Parents and grandparents claim they had no knowledge of the murder of countless people—and their children and grandchildren don't know whether or not to believe them, whether or not they *should* believe them.

But surely you must have known!

Is it possible that no knowledge of what was really going on filtered down to the ordinary Germans?

In 2011, Friedrich Kellner's diaries from the years 1939 to 1945 were published for the first time. During the war, Friedrich Kellner was a simple judicial officer. He came from a modest background and lived in the Hessian backcountry until his death in 1970. He had no access to secret files, but simply wrote down the bits of information he overheard, gleaned from conversation with other locals and, above all, read in newspapers available to the general public. His diaries are evidence of what those who "had no idea" could have known about the dictatorial regime, the war, and the Holocaust. For example, in 1941 Friedrich Kellner wrote: "The mental asylums have turned into centers for murder." Reading the newspapers, he had noticed a suspiciously high number of death notices for people in mental hospitals. He had also been told about a case where a couple was able to bring their mentally ill son home from such a hospital just in time. Around the same time, immediately after the attack on the Soviet Union, Friedrich Kellner heard about the mass murder of the Jews: "A soldier who was home on leave described the awful atrocities he had witnessed in occupied Poland. He had seen how naked Jews, men and women, were made to stand in front of a long, deep trench. Upon SS command, a number of Ukrainian men shot them in the backs of their heads and they fell into the trench. The trenches were filled in even though screams could still be heard from within!" In September 1942, two Jewish families were deported from Kellner's hometown, Laubach. In his diary he writes: "In the last few days, the Jews from our district have been deported. From Laubach they took the Strausses and the Heinemanns. A well-informed source tells me that all the Jews are being taken to Poland, where they will be murdered by SS troops."

In 1996, the artist Gunter Demnig started laying *stolpersteine*, or "stumbling blocks"—cobblestone-sized, brass memorials—in front of houses where victims of the Nazis used to live. Now in over 800 German towns and villages, they make the number of victims palpable: In some streets there are *stolpersteine* in front of every other house, sometimes with a single name, sometimes with the names of an entire family. On these streets it would have been glaringly obvious that some neighbors were missing: the Jewish family, the girl with Down syndrome, the homosexual, the communist.

Yet in many German families, the parents and grandparents have never been asked any probing questions. "The Nazis" were others. It is inconceivable that the friendly grandfather might have committed any crimes on the frontline, or that the kindly grandmother might have cheered Hitler. Just as unimaginable as it was for Jennifer Teege to discover that her grandmother once enjoyed the good life on the edge of a concentration camp.

This self-delusion, this schizophrenic view of one's own historical narrative, is rarely as apparent as it is with Ruth Irene Kalder. She was no perpetrator, but she was a bystander and a profiteer. Amon Goeth made his career, and she joined him in it. Amon Goeth remains a stranger we can distance ourselves from, but in Ruth Irene Kalder, the seduced opportunist, we can recognize some part of ourselves.

When she learned of Amon Goeth's execution from the newsreel, Ruth Irene Kalder is said to have ranted and raved. Monika Goeth remembers her grandmother, Agnes Kalder, claiming that Ruth Irene's hair went white and that she subsequently died it black.

Monika Goeth also recounts that her mother repeatedly watched the American film *I Want to Live* with Susan Hayward in the title role. The film makes a passionate plea against the death penalty; it shows an innocent woman being executed for a murder she didn't commit.

The Third Man was also among her favorite films. Ruth Irene Kalder is said to have seen herself in the beautiful actress Alida Valli. In the famous post-war film, Alida Valli plays the girlfriend of murderer Harry Lime, played by Orson Welles. She stands by her lover with total devotion, loyal to the grave.

According to her daughter Monika, Ruth Irene Kalder did go on to have relationships with other men, but she didn't love any of them as she had loved Amon. After the war, she dated a US army officer for a short while. He paid for her English lessons. Even after he returned home to his wife and child in Texas, he continued to send her regular love letters and monthly checks until her suicide in 1983.

In 1948, two years after Amon Goeth's execution, Ruth Irene Kalder asked the American authorities in US-occupied Germany to allow her to take on Goeth's name, claiming that it was only the confusion at the end of the war that had prevented them from getting married.

Goeth's father, Amon Franz Goeth, with whom Kalder had been corresponding, supported her request. He confirmed that his son and Ruth Irene Kalder had gotten engaged before the end of the war. Since Goeth's divorce from his second wife had already been finalized, Ruth Irene was allowed to drop her maiden name, Kalder. From then on she was known as Ruth Irene Goeth.

Amon Goeth lived on in her stories—as a charming, witty gentleman from Vienna who sadly died a hero's death in the war. Ruth Irene Goeth never spoke of the crimes committed during the war; in this she was no different from most of her contemporaries. More than once, Ruth Irene struck her daughter Monika because she would not stop asking questions.

Monika Goeth describes her mother as self-absorbed and cold-hearted, a woman who was mainly concerned with her own beauty. When she had a facelift, she also had her nose, which she considered "Jewish-looking," straightened. She was a woman who carried a lifelong unhappiness because the world had taken her great love away from her too soon.

It does not appear as if Ruth Irene secreted away any of the riches Amon Goeth had amassed in Płaszów. Ruth Irene Kalder worked as a secretary and sometimes posed as a fashion model for catalogues; in the evenings she often worked at the *Gruene Gans*—the Green Goose, a bar in the trendy quarter of Schwabing. Her daughter Monika has described how her mother loved strutting around Schwabing in a dress to match her lipstick, with her equally well-groomed poodle, Monsieur, by her side.

Ruth Irene was not interested in Monika or her problems, Monika Goeth has said of her mother: "Ruth wasn't concerned with much. What feelings she had were reserved for her dead lover, for Amon."

*   *   *

THE BOOK ABOUT MY MOTHER delivers a double blow: It broke my enchantment with my grandmother twice. She is described as heartless and selfish: first at the side of the concentration camp commandant, and then as a terrible mother. Even worse than that—as a monstrous kind of mother who neglects and hits her daughter. In the book, my mother criticizes her severely; she hardly has anything kind to say about her at all.

I don't think that is very fair to my grandmother. After all, she is dead and can't defend herself.

At the same time, the book clearly illustrates the degree to which my mother's life has been dominated by her struggle with my grandmother. But despite everything, they always kept in touch. When my mother was pregnant with me, she even lived with Irene for a while.

What is striking is the close relationship my mother had with her grandmother Agnes, Irene's mother. My mother spent her entire childhood living with her mother and grandmother in the apartment in Schwabing. A household with three women, three generations beneath one roof. The men—my great-grandfather and Amon Goeth—were dead.

According to the book, my grandmother was jealous of Monika's close relationship with Agnes; she felt like an outsider between two soulmates. During my mother's childhood, Agnes was her calm anchor and her pillar of support.

Sometimes I feel like history is repeating itself: Just as my mother had a close relationship with her grandmother and a difficult one with her mother, I used to feel at home with my grandmother and uncomfortable with my own mother. It appears that love tends to skip a generation in my family.

Again and again my mother stresses how obsessed my grandmother was with her looks, and how beautiful she was—like a young Elizabeth Taylor. While Irene was always dressed smartly, Monika allegedly ran around in rags.

Irene is reduced to her failures as a mother, and to her vanity—as if she had spent half her life in front of the mirror perfecting her makeup.

I don't believe that Irene was only selfish and vain. She was an attractive and unusual woman. She didn't look for a provider, which was the norm in postwar Germany, but stood on her own two feet: She worked as a secretary at the Goethe Institute for many years. This is something else we have in common: I also worked at the Goethe Institute, while studying in Israel.

What was unusual for somebody of her generation was that she spoke very good English and often read the British *Times*. Her apartment was full of books, among them Tucholsky, Boell, and Brecht. She was interested in drama and literature. She voted for the socialist SPD party and was a fan of the politician Willy Brandt.

My grandmother was very liberal for her time: For a while she shared her flat with a transvestite called Lulu and went out on the town with him and his gay friends. My parents met when one of my father's friends, also African, was living as a lodger in my grandmother's house. Having an African man living in your house was far from normal in Munich in the 1960s and '70s. She was no racist.

I would have loved to ask my grandmother's friends some questions about her, but all I have is what journalists have reported about her—and my mother's opinions. Neither source is exactly full of praise. Usually I can trust my gut; I have a keen sense of character. Can I really have been so wrong about my grandmother?

When I was about seventeen years old, my adoptive parents gave me a postcard from my grandmother. They had kept it from me until then because they were worried that I would be torn between my old family and my new one. My grandmother had sent me the card for my seventh birthday, together with a picture book she had picked out for me. I would have preferred to have received these things much earlier. It would have been helpful and important to me—after all, they were tangible possessions, memories of the natural family who had suddenly disappeared from my life when I was adopted.

My grandmother's postcard is of a painting by Paula Modersohn-Becker, *Peasant Girl with Arms Folded*. It shows a serious and

Paula Modersohn-Becker's painting Peasant Girl with Arms Folded. *Ruth Irene Goeth sent this card to her granddaughter for Jennifer's seventh birthday*

proud-looking young girl, her arms crossed in front of her body. She is about the same age I would have been when my grandmother wrote the card. Irene's handwriting is itself a work of art. She took great care in writing the card, care that is typical of her generation. She wrote, "Dear Jennifer, wishing you a wonderful birthday and another 364 happy days in your new year. Do you like reading? I hope so, and if you do I am sure you will enjoy this book. I think of you often. Give my regards to your parents. Yours, Irene." The message is lovely and heartfelt. It makes me happy to see the word "Yours" in front of her signature.

■　　■　　■

Inge Sieber, Jennifer Teege's adoptive mother, remembers that for a while, when she was still small, Jennifer hoped that her grandmother would take her in.

Once, before the adoption, Ruth Irene Goeth visited Jennifer and her new family. She called beforehand and asked if she could come by, so Jennifer's foster parents invited her for afternoon coffee. Inge Sieber found Ruth Irene to be friendly and thoughtful; she was wearing a long patchwork skirt and looked nothing like a grandma. "She was dressed in these shabby-chic clothes. Flashy, extravagant, but not false. I was 25 years younger, but next to her I felt like a little old housewife." Jennifer's grandmother stayed for a couple of hours, asked lots of questions, and was generally very interested in Jennifer's new family.

At around the same time, in the mid-'70s, the Israeli historian Tom Segev visited Ruth Irene Goeth at her apartment in Schwabing.

Segev was not yet the scholar and journalist of international fame he would go on to become; he was just a young PhD student from Boston University. For his thesis on concentration camp commandants, he traveled all over Germany to interview the close relatives and friends of many Nazi perpetrators. He hoped that they would provide answers to his questions about the camp commandants' frame of mind and the motives that drove them. His work was published under the title *Soldiers of Evil: The Commandants of the Nazi Concentration Camps.*

His study sheds light on not only the commandants' psyche, but also that of their closest relatives, usually their widows. Segev writes: "They agree to be interviewed by me because they are haunted by a past that they cannot seem to escape. . . . Each of these people hoped that they could whitewash their past, if only a little bit."

All of those questioned played down their time at the camps. For example, Fanny Fritzsch, widow of Auschwitz prison camp leader Karl Fritzsch, had "no difficulties to explain the atrocities her husband was accused of. She had simply decided that they had never happened," Segev writes. According to Fanny Fritzsch, nobody died at Auschwitz. She told Segev that Fritzsch had been "the best husband on Earth," and that she had raised their children based on his model.

Ruth Irene Goeth stands out amongst the interviewees: True, she also plays down what Amon Goeth did, but at the same time she puts on a show as the profligate widow, downright wallowing in reminiscences of her time in the camp. Segev writes about his visit with Ruth Irene Goeth:

"In the late seventies, Ruth Goeth lived in an apartment that had seen better days. She received me in a Chinese wrap dress. Dark green velvet curtains covered in dust and heavy furniture gave the apartment a gloomy air. She sat down on the sofa, crossed her legs and—through a long holder—smoked one cigarette after another, her little finger coyly raised. It was a carefully staged performance; it was not difficult for the former actress to affect a nihilism that was evocative of Weimar Germany in the 1920s. 'Oh, well, Płaszów,' she said in a husky, raspy voice. 'Yeah, yeah, Płaszów.' She paused and then suddenly said: 'They will tell you that I had a horse there and that I was a whore. Indeed, I did keep company with a large number of officers, but only until I met Goeth. And he gave me a horse. I used to love riding then. Ah, Goeth—a picture of a man!' I could not help feeling that she enjoyed every single moment of her performance. . . .

"'It was a wonderful time', his widow said. 'We enjoyed each other's company. My Amon was king, I was his queen. Who wouldn't have relished that?' She added that she was only sorry it was all over."

As to Amon Goeth's victims, Ruth Irene Goeth adds: "They weren't really people like us. They were so filthy."

■   ■   ■

I AM STANDING AT THE TOP of the watchtower, looking out over the vast area of the Auschwitz-Birkenau concentration camp. "The size of at least 100 soccer fields," another tourist says next to me in German.

An icy wind is blowing around the tower. I think about zipping up my coat. The people here suffered terribly from the cold. Will I be better able to share their feelings, their desperation, if I keep my jacket open? Must I keep it open? Must I imagine what it was like being crammed double and triple into bunk beds, living in drafty barracks without heating or stoves, not being allowed to go to the toilet at night, and how they dealt with it if they were suffering from diarrhea?

Is there any point at all to my coming to Auschwitz—hasn't it all been written up in the history books anyway?

I've never been here before, but if I had to draw a picture of a concentration camp it would look like this. The gate to Auschwitz-Birkenau, the tracks that lead into the camp, the vast sky above the barracks. When I think of the words "concentration camp," I see the tracks of Birkenau in my head—and the faces of the emaciated people who were liberated from the camp, their large eyes set deeply in their sockets. These images are imprinted on most people's minds—and firmly fixed in my own.

I walk along the tracks until they end abruptly. Many of the people who climbed down from the cattle cars right here were already half-dead when they arrived. At the ramp, they were divided into two groups: one went straight to the gas chambers, and the other was forced to work. The trains from Płaszów probably arrived here, too.

The gas chambers and crematoria stood at the edge of the meadow, in front of the birches that Birkenau is named for. Shortly before their withdrawal in January 1945, the Nazis dismantled the buildings and demolished the last remaining crematorium.

Over a million people died here. We are walking over their ashes.

Some of my fellow travelers are asking lots of questions; I just listen. In the children's barracks, simple pictures have been drawn on the cold, bare walls. Pictures of a normal childhood: children with a doll, a toy drum, a wooden pull-along pony. I am reminded of my own sons. These children here were alone, unprotected.

The tour guide hurries us along; we have to get back to the bus—we need to move on to Auschwitz I, the smaller main camp. A few minutes later we arrive, and I walk through the entrance gate displaying the motto "Arbeit Macht Frei"—"Work brings freedom." I recognize it immediately; I have seen it in photographs innumerable times. It feels strange walking through here now, unreal somehow.

When I visited the Płaszów memorial yesterday, I came not just as Jennifer Teege but also as the granddaughter of Amon Goeth.

My grandfather was the key figure there, so the place concerned me directly. Now, one day later in Auschwitz, I am but one visitor among many.

The tour of the fenced-in compound begins. The path takes us to the row of red brick buildings that house the museum's exhibits with their display cases, photos, and statistics. So many numbers. There is something impersonal about numbers; they confuse me. I prefer letters.

I walk from room to room, from one building to the next. I am not prepared for the room that I enter next: a huge pile of spectacles behind a glass wall. After that, a room full of shoes: boots, sandals, a lady's red slipper.

And then a mountain of human hair. Why must I suddenly recall my last haircut? Just a small amount of trimmed hair was left on the floor. Here, there are two tons. When the Red Army liberated the camp, they found seven tons of human hair, two of which are exhibited here. Seven tons of human hair, an inconceivable amount. The hair belonged to the murdered women and girls; it was destined for use in felt materials and sweaters.

More displays: crutches, prostheses, wooden legs, stilts, hairbrushes, shaving brushes. And pacifiers, baby clothes, two small wooden clogs, tiny knitted mittens.

Behind glass, suitcases with names and addresses written on them in chalk: Neubauer Gertrude, orphan. Albert Berger, Berlin. There is a Hamburg address, too.

I step into narrow walkways lined with row after row of photographs of the camp inmates. I enjoy photography, especially portraits. I like to get really close up so I don't miss anything. I study these photos closely. Some prisoners look proudly into the camera, others are frightened. Most of the faces wear vacant expressions. These are portraits of the dead.

In the beginning, the newly arrived prisoners were photographed; later, they were tattooed with identification numbers instead. The ink used to tattoo the prisoners was supplied by the Pelikan company. At school, we used to write with Pelikan fountain pens and Pelikan ink, and we never thought anything of it.

I go back outside, sit down on a bench, and take a few deep breaths of fresh air. I need a break and want to be on my own.

After a little while I rejoin my group. We visit the so-called "death block" next, hidden behind tall walls. Prisoners were shot in the yard. Nobody could see in from the outside, but the cries and the shots could be heard. I go down into the dark basement. There are narrow chambers within the walls: standing cells, too tight to sit down in. Prisoners had to crawl to get inside. After a day's work, four prisoners would have had to share one of these cells, standing up in the dark the whole night long. This was a punishment for so-called camp crimes. One prisoner, for example, was sentenced to seven nights in a standing cell for hiding a cap in his straw mattress for protection against the cold. The cells would not be opened again until the following morning. Sometimes a man would have died by then, and the other prisoners would have spent the night pressed tightly against his corpse. I push for more details. Who comes up with such cruelties? People like my grandfather. There were standing cells at Płaszów, too.

More and more visitors are pushing down into the cramped basement. People are pushing me from all sides, and I hasten back outside. In a way, it is a good thing that so many people come to Auschwitz, that they are not running away from history.

Near the office of former camp commandant Rudolf Hoess, we pass the gallows where Hoess was hanged after the war. The man who organized the mass murders at Auschwitz. I remember reading that, when Hoess and my grandfather were extradited to Poland together, the mob attacked Amon Goeth and wanted to lynch him. I was shocked. Their reaction illustrates the extent of the hatred people felt for my grandfather. It wasn't Spielberg's film that turned my grandfather into evil personified; he had already become a symbol of sadism while he was still alive.

A group of teenagers has gathered around the spot where Hoess was executed. I watch them from a distance and wonder what they are feeling right now. Anger? Gratification? Indifference?

One of the gas chambers avoided demolition, and one of the crematoria as well. The room is dark, the ceilings are low. I peer into the dark hole of the incinerator while, next to me, other tourists are filming the surreal surroundings with their cell phones.

And then it all becomes too much; I need to get away from Auschwitz. I feel as if I am being choked. This place is too dark, like a deep hole, like a grave that is pulling me in. I don't want to be sucked in. It doesn't help the victims or me if I only see myself as a perpetrator's granddaughter and only grieve and suffer as such. I'm glad I came here, but I'll never come back.

I wish someone had made my grandmother come and see this place. Surely, here, she would have had to open her eyes at last.

■　■　■

In the early 1980s, London filmmaker Jon Blair was working on a documentary about Oskar Schindler in agreement with Steven Spielberg. In doing so, Blair carried out much of the research that laid the foundation for the feature film *Schindler's List*. He talked with Schindler's widow Emilie and many survivors. Even 65-year-old Ruth Irene Goeth agreed to an interview, despite being very ill; she was suffering from pulmonary emphysema and was at times dependent on an oxygen tank. Ruth Irene Goeth expected Blair to come alone and to ask her about Oskar Schindler. Instead he rolled up with an entire film crew and asked questions about Amon Goeth. It was a long interview.

The video shows a well-groomed and made-up woman, her jet-black hair piled up high on her head, yet she is ravaged by her disease and constantly gasping for air. She speaks English and chooses her words carefully.

And she still defends Amon Goeth: "He was no brutal murderer. No more than the others. He was like everybody else in the SS. He killed a few Jews, yes, but not many. The camp was no fun park, of course."

She claims that, before the camp, Amon Goeth never had any dealings with Jews. She omits any mention of the bloody ghetto liquidations Amon Goeth organized before and during his time as commandant at Płaszów.

After the war, Ruth Irene Goeth was still in contact with Oskar Schindler, sporadically and on friendly terms. In the interview with Blair, she says that Schindler treated the Jews well, but mostly because they were useful to him. Schindler, Goeth, she herself—"we were all good Nazis," she says, "we couldn't be anything else." There had been no alternative, she claims; nobody liked the Jews—that was how they had been brought up.

When asked about herself, Ruth Irene Goeth explains: "I always felt that it was all wrong, but I wasn't the one who made the rules of those times. Whenever Amon and I had an argument, I would say that I was leaving; I didn't want to have to witness any more of this. But then the maids would come and say to me, please don't leave us, you always help us, what would we do without you?"

She describes herself as a guardian angel to those girls: "The whole camp was saying, 'God has sent us an angel.' And that angel was me."

When Blair observes that the only reason Ruth Irene Goeth had to protect the maids in the first place was because Amon Goeth would threaten and beat them, she counters that most people weren't treating their staff particularly well at the time.

In the end, it wasn't the maids' pleading that made her stay, it was her love for Amon Goeth. Ruth Irene says: "He was a very handsome man, well liked by everybody. He was obliging towards his friends, and he was charming—just not towards the prisoners, no, not at all." Amon Goeth had closer dealings with some inmates than others, and some he actually liked, she contends. But, she adds, there were so many Jews in the camp, it was impossible to know each and every one of them.

When pressed by Jon Blair, Ruth Irene Goeth admits that there were indeed old people and children in the camp, yet claims that she "never saw" the children. But then she recalls the children being deported, as already described to her daughter Monika—presumably when the trucks left Płaszów for Auschwitz. "Only once did I see that they were taking children away on a truck, and I was very sad; it tugged at my heartstrings. But a friend of mine said, 'They're only Jews.'"

When questioned by Blair whether she has any regrets about those times, she answers: "Yes, yes, honestly. But I never hurt anyone. Nobody can prove that I ever did anything wicked."

She claims that she never went inside the camp, never went near any of the barracks. She stayed in the villa, inside her "four small walls." From the villa, she was only able to see the prisoners in the quarry; they appeared to be completely normal workers, she argues. And no, she did not know that some of the workers died at the quarry. No, she never saw the executions that took place on the hill either, just a few hundred yards from her house.

Women forced to work at Płaszów

Shortly before her death, Ruth Irene Goeth showed her first signs of remorse. She told her daughter Monika: "I should have done more to help. Maybe my illness is God's punishment for not doing enough."

On January 29, 1983, the day after her interview with Jon Blair, Ruth Irene Goeth took an overdose of sleeping pills.

Perhaps Ruth Irene Goeth was afraid of what was to come once Jon Blair's documentary aired. Yet that certainly wasn't her main reason to commit suicide: She had first spoken of taking her own life months before the recording session.

In her suicide note to her daughter, Ruth Irene Goeth wrote: "Dear Monika . . . Please forgive me for all the mistakes I've made . . . I am leaving. I am a wreck. A burden to myself and everybody else. It is so hard to be locked up with this illness all by myself. I want to go to sleep and never wake up again. Everywhere I look, fear is staring back at me. Believe me, it wasn't an easy decision, but this life, being chained to the couch, is dreadful. Take care. Don't be so hard all the time. I have been so desperate. My life would have been one long illness. . . . Remember me well . . .

you didn't make it easy for me either. But I have always loved you as you love your own child. Your mother."

Not a single word about her time with Amon Goeth.

* * *

I CAN HARDLY WAIT TO SEE my grandmother in the documentary about Oskar Schindler; I haven't seen her for so long. Now I will be able to look at her on the screen. I borrow the film from the same library where I first found the book about my mother.

The film consists of a string of interviews that the director, Jon Blair, conducted with contemporary witnesses. I keep asking myself when my grandmother will make her appearance. I fast-forward and rewind but can't seem to find the segment. At last, at 17 minutes, my grandmother appears.

She sits on a chair, bolt upright, looking straight into the camera. Her face is beautiful with fine features. She still has something youthful about her. She looks just as I remember her. As if time had stood still.

What she says is of no importance—I am not listening, I'm just staring at her.

I have missed her.

Now she is struggling to breathe. I can hardly bear to watch her gasping for air. She is terminally ill. I feel sorry for her.

I rewind the film over and over and watch the scenes with my grandmother a second, third, and fourth time. Only later do I take in what she is saying. She is still defending herself. She has had time to reflect, but she is immovable. She hasn't changed.

I am sad and angry. Angry with her, but also with the people around her. They point the camera directly at my grandmother; she is obviously very ill, and yet they won't leave her in peace. Irene gives evasive answers; her English is excellent. Her voice sounds familiar, lovely. Like it used to. If only she wasn't saying these outrageous things. I hear her suppressed anger. She feels like she's been cornered.

Tomorrow she will be dead. Is she already thinking of the pills she is going to take? When did she start saving them up? Has she already got the typewriter ready for her suicide note? Or is it during the interview that she realizes the time has come, that she cannot escape her illness, or her past?

They bombard her with more and more questions. I can tell by the look on her face that she would like to be left alone, that she doesn't want to give any more answers. But the camera remains trained on her exhausted face. I watch her eyes; I like her eyes the best. They say that if you're lying, your eyes become evasive, flutter, look upward, away. But she doesn't look away. Her eyes look straight ahead. I don't catch her lying. That makes it even more painful: She really believes what she is saying.

Her suicide note doesn't mention the victims—it is all about her and her illness.

Also, her letter only mentions one child; she is referring to my mother's second daughter, my younger half-sister. I was adopted and therefore gone. Nevertheless, I believe that my grandmother was thinking about me right until the end.

I would like to have had a different grandfather. But I would choose her as my grandmother again.

Perhaps it would have been different if I had met her again when I was older. If I could have talked to her, if she had repeated the same family lies that made my mother's life so difficult. I'm sure my inner turmoil would be even worse. But I never discussed anything like that with her. I was just a child.

None of this means that I agree with what my grandmother did, or that I want to cover for her. I renounce what she did, and sadly didn't do, in the camp. I renounce her explanations. I simply distinguish between the public figure, Ruth Irene Goeth, and my grandmother, Irene.

Many children and grandchildren of Nazi criminals feel the need to apologize on behalf of their relatives. The grandchildren do this less fervently than the "penitent generation" of Nazi children. At first, I noticed the same response in myself, a fear of how my environment might react if I admitted that I love my grandmother.

The book about my mother is entitled *I Have to Love My Father, Don't I?* It implies that actually she mustn't love her father. The author, Matthias Kessler, keeps confronting my mother with her father's crimes. I can see him wagging his finger on the pages, can hear his judgmental undertone: You must not love this father, this monster!

But human psychology doesn't work like that. If my children were ever to commit some horrendous crime, I would condemn their actions resolutely—but I would never stop loving them.

The descendants of perpetrators are often accused of deluding themselves by constructing a psychological framework to help them deal with their ancestors' terrible crimes and to save the image of a good father or a good grandmother. Allegedly, they convince themselves that the offending relative was just acting on orders, that others committed atrocities of even greater magnitude. The descendants are said to cling to a fantasy that the culprit repented at the last minute.

I don't think this is true. I did wonder whether Amon Goeth might have eventually regretted his actions, but not in order to exonerate him—and therefore myself. It was much more because I am interested from a psychological point of view, and because I see

Amon Goeth as a human being after all. It is in our nature to have compassion.

My grandfather did not repent in the end; why else would he have raised his arm in a Hitler salute at the gallows? My grandmother never really repented either. She never really saw the victims; she walked through life with her eyes closed.

Nonetheless, I still feel close to my grandmother. I will not try to justify the fact; I won't explain it either. That's just the way it is.

When I was a little girl, she made me feel that I wasn't alone. I will always remember her for that.

Ruth Irene Goeth in 1983, shortly before her suicide

CHAPTER 4

LIVING WITH THE DEAD:

MY MOTHER, MONIKA GOETH

This man, he dominated our entire lives.
Even though Amon Goeth was dead, he was still there.
—MONIKA GOETH IN 2002

WANT TO SEE MY MOTHER AGAIN, not for a reckoning, but for answers to all the questions I have going around in my head.

Almost twenty years have passed since we last met. Will she be surprised to hear from me now? Will she be shocked? Curious? Pleased?

Will she like me, or reject me?

I set myself a deadline; I wanted to contact my mother before the end of the year. It was October when I went to Krakow. November came and went. Now we're in December, and the weeks are quickly passing by.

By mid-December I can't run from it any longer. A week before Christmas, I write a letter to my mother. I have found her address in the phone book.

I stumble at the first hurdle. How should I address her? Mother? Monika? Dear Monika?

As a child I called her *Mama*, but that was a long time ago.

I start my letter with "Dear Monika" and wish her a merry Christmas. I tell her that I would love to see her in the new year. I give her my address and include a small photo of my husband and my two sons. I am sharing a small part of my life with her so she'll know a few things about me now: that I live in Hamburg and am married with two children.

I know so little about her, and yet so much—from the book about her, the documentary, and the Internet.

That was one of the reasons I went to Krakow—to be closer to my mother. She had gone there herself, tracing the steps of her own parents. It was there that she met with Helen Rosenzweig, Amon Goeth's former maid. In the documentary about that meeting my mother looks lost and lonely.

The journey to Poland was an important step for me; I felt much better afterward. The moment I laid flowers at the Płaszów memorial, I felt as if a huge weight was lifted from my shoulders. Since then, I have stopped busying myself with historical facts. When I close my eyes I no longer see my murdering grandfather before me, nor my grandmother gasping for air. I am able to devote myself back to my own life and my children.

By being able to take a step back, I may have achieved what my mother never managed to do.

I study my mother's photographs again. She looks so sad in some of them, her expression veiled, that it frightens me. I'm afraid these photos could pull me back into the depths of depression.

But without talking to my mother, I will never truly understand my family and my past. There are details only she can tell me.

At first, I wasn't sure I really wanted to see her again. After I found the book, all I wanted to do was shout at her and tell her how disappointed I was. How could she leave me, and then pretend that I didn't exist? How could she think it wasn't necessary to tell me the shocking details of our family history?

But I knew it would have been futile to contact her feeling like that. I would have confronted her in the role of a small, hurt child, unable to have a proper conversation.

I needed time to understand her better.

So I kept quiet. Only during my therapy sessions did I really allow my anger to boil up. My husband suggested I confront my mother sooner rather than later. He was angry with her, too, I could tell—maybe even more so than I.

Yet I considered it best to wait. I thought about my mother a lot. What kind of person is she? Difficult, an enigma. I pored over the photos I have of her, and I watched the documentary about her and Helen Rosenzweig with new eyes. She has a peculiar gait; her shoulders are hunched as if under a heavy burden. Now, I feel sorry for her. My anger has abated a little.

Looking at her life now, I can see why she didn't feel capable of bringing me up. I can even understand why she wouldn't talk about the past for so long.

Still I would like to hear it from her: Why did she give me away? And I want to know how she has been all these years.

I no longer see her simply as a mother who abandoned her child; now I see her as the daughter of Amon Goeth—the father who became the story of her life, the story that shaped her identity. That occupied her to such an extent that it left no room for other people, for her role as a mother, or for me.

■ ■ ■

Monika Goeth was born in Bad Toelz in Bavaria. Amon Goeth had arrived there in the confusion of the last days of the war, but he was soon arrested by the recently arrived American forces in May 1945. Ruth Irene Kalder followed him to Bavaria and gave birth to their daughter in November 1945.

While in prison, Amon Goeth wrote to Ruth Irene Kalder: "Dearest Ruth, at last we are allowed to write home. I have been thinking of you, especially in early November. Was it difficult, you poor thing? What is it—a boy or a girl? Hoping to see you . . ."

After the birth, Ruth Irene Kalder came down with scarlet fever, and Ruth's mother Agnes looked after her newborn granddaughter. For the first few weeks, due to the risk of infection, Ruth was only allowed to see her daughter behind glass. The distance between mother and daughter would always remain: Agnes became the person Monika would most closely relate to, she was the girl's "Oma." Her mother would always be simply "Ruth" or "Irene."

Much later, Monika Goeth talked about her childhood at length in an interview with Matthias Kessler, the author and documentary filmmaker.

When Monika was six months old, her mother was walking her in her baby carriage when a man suddenly attacked the carriage and stabbed the baby with a knife. Monika needed an operation; she was left with a scar on her neck. Ruth Irene Kalder assumed that it must have been a former inmate of Płaszów who tried to kill her daughter.

When Monika was ten months old, her father was hanged in Krakow.

That didn't stop Ruth Irene Kalder, who changed her name to Goeth in 1948, from romanticizing about her Amon, or "Mony" as he was called. Monika, also called "Moni," was in Frankfurt with her mother when she met with an old friend, Oskar Schindler. He took one look at the girl and exclaimed: "Just like her father!" Monika was pleased to hear it.

She wasn't suspicious until the day her mother shouted at her, during an argument, "You are just like your father, and one day you will end up like your father!" That was when Monika, now twelve years old, began asking questions. Her father had died in the war, hadn't he?

Her mother wouldn't explain anything. Eventually Monika managed to press her grandmother Agnes for the truth: "Well, they hanged him, you know. In Poland, where they had killed the Jews. Your father was one of them."

Jews? Monika didn't know any Jews. By then, the family had moved to Schwabing—the bohemian quarter of Munich—but even there in the big city Monika had heard nothing about the Holocaust. She would later describe the postwar atmosphere in the 1950s and '60s as a time when "people wouldn't talk about the Jews. They were extinct like the dinosaurs."

Monika pressed her grandmother for more information: "And where was Irene?" She replied: "Also in Poland."

The following day, Monika approached her schoolteacher and asked what had really happened with the Jews. The teacher told her to concentrate on her math homework instead of worrying about things that didn't concern her.

So Monika went back to her mother and probed for answers: How many Jews did her father kill, and why? Had he killed children, too?

When Monika wouldn't stop asking questions, her mother struck her.

Ruth Irene Goeth would downplay Goeth's actions and repeat the same story over and over: The camp Mony led was only a labor camp, not an extermination camp. Another commandant, Rudolf Hoess, had been in charge of a much bigger and much worse camp in Auschwitz. Ruth Irene Goeth assured her: There had been no children in Płaszów; she never saw a single child.

Ruth Irene also claimed that Amon Goeth had shot only a few Jews, and then only for "hygienic reasons." Monika Goeth quotes her mother as saying, "The Jews never went to the toilet, which caused diseases to spread. And when Amon once spotted a few men not using the toilet, he shot them."

Monika Goeth grew up surrounded by lies. She was a child, so she believed her mother. The words we hear as children take hold, and they keep flashing through our minds whether we want them to or not.

It would take Monika Goeth almost half a lifetime to find out the whole truth about her father and her family. She went on to read all the material she could find to help uncover the lies and half-truths in her head, a process that nearly drove her insane.

It took a lot of energy to tear down the web of lies her mother had created. It would have been easier not to, especially since the stories her mother had told her were feel-good stories—stories about a charming, loving, and witty father who ultimately hadn't done anything wrong. Looking back, Monika Goeth said, "I used to view my father as a victim—a victim of the Nazis, of Hitler, of Himmler."

Psychologist Peter Bruendl in Munich explains that, to ensure normal development, children need to grow up thinking, *My parents are good people:* "It is awful to have murderers for parents, to have to think, *me, a child of killers.* That's why many people accept their parents' silence on the subject, and they too keep quiet about it. They don't ask questions about what their parents actually did during the war."

The generation born in the years just before and after the end of the Third Reich came to realize that their parents would refuse to engage in any discussion of the Nazi years. Regardless of whether they had been SS men or common Wehrmacht soldiers, most fathers (and their wives or widows) would say little to nothing about the years before 1945. They left it to their children and grandchildren to eventually rediscover their family history. With the truth cloaked in silence, legends and prejudices flourished.

Agnes Kalder asked her granddaughter to let it rest, but Monika kept needling her mother. If Ruth Irene asked her daughter to clean the bathroom, Monika would reply, "I'm not your maid from Płaszów!" If her mother hit her, she would yell, "Don't stop! Come on, hit me again, you're just like the old man! It's not me who's like him, it's you!"

When Monika Goeth was in her twenties, she became friendly with a bartender at the Bungalow, her local bar. One day, when he rolled up his sleeves before doing the washing-up, he revealed a number tattooed on his forearm. Monika was aghast. "Manfred," she asked, "Are you Jewish? Were you sent to a concentration camp?" "Yes," he replied tersely. Monika wanted to know where he had been sent. Her questions made him uneasy, but he eventually admitted that he had been in Płaszów most of the time. Monika Goeth was relieved: "Goodness Manfred, am I glad that you were only in a labor camp and not in a proper concentration camp. You will know my father then; he was Amon Goeth."

When the bartender grasped what Monika had said, he turned pale. Much later, Monika Goeth would say that she could still hear his screams: "That murderer, that bastard!" Monika didn't understand: "But Manfred, you weren't in a concentration camp, it was only a labor camp, remember?" He didn't reply; he just stood there quivering. He wouldn't talk to her for days.

Monika Goeth urged her mother to meet her traumatized friend. Ruth Irene Goeth eventually agreed, but she wouldn't tell Monika much about their meeting. She only reported that Manfred had asked her again and again: "Why did you all do it?"

When she was barely 24 years old, Monika Goeth fell in love with a black student from Nigeria, a friend of one her mother's lodgers. Monika has described him as good looking, kind of a Harry Belafonte type. She moved in with him for a while, but the relationship was not to last. On June 29, 1970, Monika Goeth gave birth to a daughter; she called her Jennifer. Jennifer was given her mother's surname: Goeth.

At the time, Monika Goeth was working as a secretary, six days a week. She was also suffering from repeated episodes of mental illness.

When her daughter was four weeks old, Monika Goeth took Jennifer to Salberg House, a Catholic home for infants run by nuns just outside of Munich.

IT HAS BEEN THREE WEEKS since I sent the letter to my mother, and I still haven't heard back from her. I am worried that she might not get back to me at all. Maybe she doesn't want to get in touch.

In fact, that was another reason why I waited so long to contact her. I wanted to be strong enough to endure her silence.

The silence feels familiar. After my adoption she was suddenly out of the picture; I heard nothing from her, and I couldn't ask her anything anymore. Now, I try to stay calm. It took me so long to write my letter; maybe she needs time for hers.

Then, one Thursday, there is a call at my office. I am not there, so the caller leaves a message: Herr Such-and-Such would like to be called back. It was Dieter, my mother's second husband; he is about the same age as her. The last time I saw my mother, in my early twenties, she had him in tow. She had brought him along without asking me first. I would much rather have met with my mother alone.

And now it's Dieter who has gotten in touch, not my mother. I wonder why she didn't call herself. Is she sending him ahead?

The next day I call Dieter back. He tells me that he tried calling me at home but that I hadn't been in. We talk for a while until he asks me bluntly, "Why don't you simply pick up the phone and call your mother?"

Simply pick up the phone? For me, nothing is simple when it comes to my mother.

Still, my mind is already made up: I am going to call her. I need clarity at last, and I don't want to wait any longer. On Saturday my husband and the boys are out, so the house is quiet. I dial the area code, then her number. It's only a few digits, since she lives in a village.

I am nervous. It rings, once, twice, three times before she picks up. She says hello and tells me that she was very happy to receive my letter. It sounds as if she has been expecting my call.

Her voice is instantly familiar; it takes me right back to the days and weekends of my childhood when I visited her.

I like listening to her. I like the way she talks. She articulates very clearly and with long pauses. In public this can sometimes come across as theatrical.

Today I can hear the joy in her voice, but she also sounds nervous. I wonder where she is right now. I know that she lives in a single-family home. In the film about her meeting with Helen Rosenzweig, some of the scenes were filmed there.

Is she in the living room, in the hallway? Is she walking through the house with the phone in her hand? Or has she fled outside, to the fresh air? I cannot imagine her sitting on a chair somewhere; she is too impulsive for that. I am sure she'll want to be walking around. She's always been restless.

It used to make me nervous when I was a child. Her presence always brought tension to the air. I never knew what to expect from her next, which frightened me. She didn't speak much; in fact, the opposite was true—silence was her favorite form of punishment.

But today, on the phone, there's no stopping her. My mother is not at all surprised that I suddenly know about our family history. She assumes that I know about lots of things; she takes them for granted, jumping from one detail of her life to the next. In the end I ask her cautiously: "Can I come and see you? Would that be OK?" She replies without hesitation: "Of course!" When I ask her when would be a good time, she says, "For you, anytime."

When I put the receiver down, I am relieved. Given how long it took me to prepare for this talk with my mother, the call went well. She sounded happy and not hostile—that is more than I had expected.

We have agreed to meet up in February; I will go to see her in her hometown in Bavaria.

First, I travel to Munich. My adoptive parents have offered to take my sons skiing for three days, and I am staying at their house in Waldtrudering, to prepare myself for the meeting with my mother in peace and quiet.

Salberg House, the children's home where I spent the first three years of my life, is in Putzbrunn, a suburb of Munich, and only a few miles from Waldtrudering. I have driven past it on many occasions. Sometimes, I would stop outside, roll down the window, and contemplate the three-story, flat-roofed red building.

The house backs onto the woods. In front is a large yard with playground toys: a wooden ship for climbing on, a swinging bridge, a water pump. Today, a number of small children are outside playing and singing. Did we have these toys when I was here? I don't think so, everything looks brand new. As an adult I have only ever looked on from the outside, but today I'm going in.

■ ■ ■

Salberg House, a home for infants near Munich, was built in the 1960s. Until 1987, it was run by the Sisters of St. Francis, who lived in accommodations across the road. Compared to other homes, Salberg House had a good reputation. Government inspectors attested that the home was "well-managed." One inspection report further confirms that "the infants were evidently being well-fed and cared for. They all have a healthy complexion . . . The happy atmosphere in the home warrants a healthy development for the children."

These days, Salberg House predominantly looks after children who are in urgent need of a place to stay. Their parents are often incapable of looking after them; some have mental health problems, are addicted to drugs or alcohol, or are in trouble with the law. Some of the children have suffered physical or sexual abuse in their families. The children are brought here by CPS or the police.

In the seventies, it was normally the parents themselves who brought their children. Often it was single or working mothers who would ask Salberg House for help.

Back then, there was no such thing as parental leave in Germany: A few weeks after giving birth, mothers would have to report back to work; their jobs were rarely saved for longer. Many women would work full-time, six days a week. Part-time positions and childcare facilities were few and far between, and little assistance was available to mothers. Children were put up for adoption much more often than they are today.

Jennifer Goeth's case was no different. The reason for her admission to Salberg House in the summer of 1970 is listed in her files as "working mother."

In the early 1970s, up to 200 infants and toddlers were living in the home in Putzbrunn. They were divided into groups of ten to twelve children, each led by one or two nuns. The younger ones lived in the infant ward, the older ones in the toddler ward.

Today the groups have names such as "Bears," "Grasshoppers," and "Seven Dwarves." In the seventies, they were simply numbered. Today, the babies are walked about in baby carriages; back then, their cribs would have simply been pushed onto the balconies for fresh air. Wolfgang Pretzer, the current director at Salberg House, explains: "Even if it was a very good orphanage for its time, by today's standards it was more about the provision of daily essentials than actual care. In those days, they had less time to meet each child's individual needs. The groups were larger, and they had fewer care staff than we do today."

■   ■   ■

MY EARLIEST MEMORY is of lying on the floor, screaming in the dark—I must have fallen out of my crib. The night nurse comes and picks me up and returns me to my bed. Tucked under my blanket, I go back to sleep.

We used to sleep in cribs with white bars; the sides could be lowered for access. One of the sisters must have forgotten to secure the bars after putting me to bed.

When I became pregnant and had babies of my own, I often thought back to the orphanage. For nine months, each of my sons was in my womb, safe and warm. Once they were newborns, I carried them around everywhere; I would sing to them and rock them to sleep. The close physical connection they had to me in the womb continued after they were born.

It was different for me; my mother just disappeared after my birth.

The photos I have from my time at the orphanage don't show it. I look happy in all of them.

The entrance hall of Salberg House looks friendly; the walls are plastered with colorful pictures drawn and painted by the children. I called in advance to explain that I used to live at the home more than three decades ago, that I would love to have a look around. The director and a female social worker who was there during the seventies have offered to guide me.

Despite the large number of children, it is quiet. We walk down long corridors and are met by a few toddlers coming toward us on tricycles and ride-on cars.

My group of children had been led by Sister Magdalena. My adoptive mother has told me about her, how lovely and approachable she was. The older woman remembers: "Sister Magdalena's group used to be in the room on the left by the stairs—where the Bears are now."

I am allowed to enter one of the groups' living rooms. First we have to ring the bell, just like at a normal home, before one of the care workers opens the door. There is a dining room with an open kitchen and a sunny, brightly furnished living room. Further along there are three bedrooms for two to three children each—the same size group as during my time here.

A little dark-haired girl with a pale face and rings under her eyes comes toward me. She shoots me a quick glance. I later learn from her caregiver that she hasn't said a word since she arrived. There are two dark-skinned girls, too, their frizzy curls sticking up; they are laughing.

I wonder what these children have been through. Do they miss their parents? Do they want to return to their families?

During my time at Salberg House, we had designated visiting hours on the weekends. Every Sunday, as the other children's moms and dads were arriving, I would look longingly at the door: Would my mother be coming today?

※　※　※

Monika Goeth would visit Jennifer only once in a while, and even then she did not always have time for the little girl. She had married a man who often beat her. Once, outside the orphanage, he beat her so hard that she

Jennifer Teege's christening in the orphanage's own chapel; the young Sister Magdalena is her godmother

needed hospital treatment. Jennifer got to know her mother's husband during her visits. Monika Goeth would later say about him: "My first husband was just like Amon. I must have chosen him to punish myself."

Sometimes, however, Monika would take Jennifer to Ruth Irene Goeth's apartment in Schwabing.

Jennifer was christened on March 21, 1971, in the chapel annexed to Salberg House. Her mother did not attend; Sister Magdalena was Jennifer's godmother.

■　■　■

I STEP INTO THE PLAIN LITTLE CHAPEL where I was christened. I ask my guides for a moment on my own and sit down in a pew.

In the living room of my adoptive family's home there is a table with a deep drawer. We children used to keep our photo albums in it. Among others, my photo album contains pictures of my christening, stuck in carefully by my adoptive mother. A young, blonde nun is holding me over the font—Sister Magdalena, my care worker and godmother. She is wearing the white habit of the order working at the orphanage. Next to her is the priest, pouring the baptismal water over my head. In another photo, Sister Magdalena is carrying me in her arms. I am wrapped in a long, white christening robe, my tiny dark hand gripping hers. In her floor-length habit and wimple Sister Magdalena looks like a Madonna.

I believe that she and her assistants did their utmost to ensure that, even though we were in an orphanage, we children would experience love and tenderness. She strove to be a kind of substitute mother for eleven young children. At bedtime she would pray with us in the dormitory.

I would love to meet with her, but she no longer lives in the convent. She once wrote a letter to my adoptive parents and said that she had left the order. She also mentioned that she had once caught sight of me and my new family in downtown Munich. She hadn't wanted to intrude at the time but hoped that I was still doing well.

With help from the Order of St. Francis, I manage to get ahold of Sister Magdalena's email address. I write to her and receive a prompt reply. She begins her email with "Dear Mrs. Teege—or dear Jenny?"

Sister Magdalena remembers me well. She still has many photographs of me, she writes. I should come and visit her; she and her husband don't live far from Munich.

I have no problems finding their detached home in the middle-class suburb. Sister Magdalena's hair is white now; she wears it in little curls. She welcomes me with a hug.

A plain cross hangs on the wall above the kitchen door; it catches my eye as I come in. God still plays an important role in her life, she explains, but the church much less so. After she left the order she got married and had children. Now she has grandchildren, too. Her husband joins us at the table. He used to be a priest, speaks several languages, and is originally from somewhere near Krakow. He knows Płaszów, and he has heard of Amon Goeth. As I tell them about my recent discoveries, they both listen intently.

Sister Magdalena has no memories of my mother or my grandmother. But she remembers that I used to be sad when nobody came to collect me on the weekends. Some children saw their parents frequently. She tells me that I had a friend in my group whose parents came to visit her every Sunday, and that I was keenly aware of it. In the early days my mother came regularly, she says, but as time went on her visits became more sporadic.

Magdalena was in her late twenties when she was my care worker. Now she is in her late sixties, but she still recalls many details. She says that I was a happy, open, uncomplicated child and very popular in my group. She had a personal relationship with each of her charges, and she is still in touch with some of them today. Very few of the children from the orphanage have had a straightforward life, she says; many are struggling.

She shows me her photo albums: Sister Magdalena and us at the zoo; Sister Magdalena and Santa at Salberg House. There was another dark-skinned girl in my group, as well as a number of children with physical disabilities; one was blind in one eye, another was missing a leg.

Magdalena says that her job was to give an "extra portion of love." She confides that she found it incredibly hard to let go of the children when their time had come to leave the home.

My reunion with Sister Magdalena is very happy and joyful. We talk and talk; I don't want to leave.

On my way back to my adoptive parents' house, I try to remember what it was like for me to suddenly find myself separated from Sister Magdalena. She was the person I felt closest to at the orphanage. One fateful day, I was taken in by a foster family, my future adoptive family, never to see Sister Magdalena again. Did I miss her? My adoptive parents tell me that in the early days I used to talk about her a lot.

■   ■   ■

The children of Salberg House usually left by the time they were three or four years old. By then, they were supposed to have been reunited with their natural families or placed in foster care. If they hadn't been, a different children's home would be found for them.

On the weekends, prospective parents would often come to Salberg House to choose a child to potentially foster or adopt. Cute little babies were the easiest to place. Jennifer was over three years old, and her skin was dark. "Back then, it was more difficult for black children. We wouldn't even have considered placing them somewhere in the country; it wouldn't have been fair to them," a former staff member recalls.

The first family Jennifer was introduced to already had a little girl and were considering a foster child of the same age. But when they saw Jennifer towering inches over the other three-year-olds, they decided against her: Jennifer was too tall for them.

At about the same time, a professional couple from Waltrudering, near

Munich, applied to foster a child: Inge and Gerhard Sieber. Inge is from Vienna and has a PhD in education; Gerhard, an economist, is from Bochum. They had had two sons in quick succession who were now three and four years old. They had been difficult births; both boys were born early.

Since they had always wanted three children, Gerhard Sieber suggested to his wife that they take in a foster child. This was nothing out of the ordinary for him; his sisters and his mother, who eventually became Oma Bochum to Jennifer, had given many foster children a temporary home. Gerhard Sieber considered helping children in need a lovely family tradition.

Inge Sieber speaks with a trace of a Viennese accent as she recalls her feelings at the time: "I was less sure than my husband. I was afraid that we might find ourselves with an emotionally damaged child and that I might not be able to cope."

In 1973, despite her concerns and with her two young sons in tow, Inge Sieber went to her local fostering authority and applied to foster a child. At the time, adoption was not on the table; all they wanted was to help a child for as long as it took. Inge Sieber explains: "To our minds, adoption was something for people who couldn't have children themselves. We already had two sons and didn't want to deprive a childless couple of the opportunity to adopt a child of their own.

"At the appointment, my two little boys were being so wild that I was convinced the agency would never allow us to foster a child. I was sure they were thinking, 'This mother can't even control her own children.'"

Yet the agency considered the Siebers suitable. A social worker visited the family in their home, and Inge Sieber had to undergo a health check. In those days, it was predominantly the future mothers who were checked, as it was assumed that the child would only be cared for by the woman. The Siebers' case was no different: Inge Sieber was a housewife and looked after her children; in her spare time, she volunteered in the neighborhood, supported the elderly in her community, and gave private tutoring in Latin.

Three months later, the Siebers received a phone call: "There is a mixed-race girl in a children's home in Putzbrunn who is in desperate need of a foster home."

Today it would be unthinkable, but the Siebers were put in touch with Salberg House without counseling or support of any kind. At the time, the fostering and adoption process was often railroaded through. Commenting on the practices of the early seventies, the chronicles of Salberg House state: "Prospective adoptive and foster parents would often show up at the door unannounced, proffering a letter from CPS that authorized them to inspect a certain child and to take the child home there and then. It took time and effort for the realization to take hold that it was in everybody's best interest for them to first spend some time getting to know each other."

Inge and Gerhard Sieber talked about the idea of fostering a child with their young sons. Matthias, the older of the two, remembers that his parents told them: "There is this little girl, and we're going to have a look at her."

When the Sieber family went to visit Jennifer for the first time, the boys had a picture book for her and a blue teddy bear. Inge Sieber recounts: "We saw a happy little girl with wild, spiky hair—the hair close to her head was growing in natural curls, but her mother had straightened the rest of her hair, so it stood up on end in spikes. Jenny was effectively presented to us—like some merchandise."

Another girl from the children's home immediately climbed onto Inge Sieber's lap, looked at her, and declared: "Mama, you are nice." Inge Sieber remembers how sad it made her that, to this little girl, a "mama" was any female visitor.

The Siebers took Jennifer for a walk and visited her a few more times after that. Eventually, Jennifer went for a "trial day" at the Siebers' home in Waldtrudering. For lunch, Inge Sieber served chicken. She says: "It appeared that Jenny was used to softer children's fare. She was surprised at the bones, picked at her food, and chewed the same mouthful over and over. I asked her, 'Don't you like it?' and Jenny replied: 'No, I don't eat cat!'"

For her afternoon nap, she slept in Matthias' bed; he temporarily moved to the guest bed to make space for her. Jennifer was a friendly and straightforward little girl, and she seemed to feel comfortable with the Siebers. At the end of the day, before they took Jennifer back to Salberg House, they asked the three-year-old, "Do you want to come live with us?" Jennifer said yes.

Inge Sieber went shopping with her sons to prepare for Jennifer's arrival. She bought a milk mug and asked Matthias and Manuel who it might be for. "For the little girl who is going to be our new sister soon," the boys replied.

On October 22, 1973, Inge Sieber collected Jennifer from Salberg House for the last time—to take her home. She was given Jennifer's health record with details of her vaccinations and which childhood diseases she'd had. Sister Magdalena gave her a collection of photographs. But one thing Jennifer did not have after three years in a children's home was a favorite stuffed animal or security blanket of her own.

Inge Sieber remembers: "The first thing I did was to take Jenny for a little walk. We went to the butcher's, and when he handed her a slice of pepperoni she beamed at him."

Frau Sieber was amazed at how "happy and mature" Jennifer was. She had been expecting a shy, traumatized, institutionalized child, but "Jenny was more confident and more independent than my own sons. She knew her way around everyday life. The social workers at Salberg House had prepared her well; they had taken the children grocery shopping in the village, for example."

One thing was unusual, however: At first, Jennifer would not leave Inge Sieber's side. She followed her foster mother wherever she went, even to the bathroom.

According to Inge Sieber, Jennifer was an inquisitive child, hungry for knowledge. When she saw Matthias and Manuel's toys, she asked: "Whose are those?" Inge Sieber replied: "They belong to all three of you."

Jennifer's older brother Matthias says that he and his brother were excited about their new playmate and that they liked her instantly. They never felt any jealousy toward their new sister, he adds.

Gerhard Sieber built triple bunk beds for the children. Jennifer slept in the bottom bunk; Manuel, who was about the same age as her, slept in the middle, and Matthias, who was one year older, slept on top. A photo from those early days shows all three of them laughing in their bunks: the two small, blond Sieber boys in red-and-blue striped pajamas, and thin, tall Jennifer in a nightdress made from the same material.

Every year on October 22, the anniversary of Jennifer's arrival in Waldtrudering, the family would give her a little present. "October twenty-second was something like Jennifer's name day for us," Inge Sieber explains.

■ ■ ■

I LOVE THAT PHOTO of the three of us in matching nightclothes, on the bunk beds.

After Inge and Gerhard had put us to bed, we would have our stuffed animals and dollies talk to each other for a while: Manuel's teddy—Grizzly—would growl, Matthias's teddy Rascal would interrupt, and Jimmy, my dark-skinned doll, would also chime in.

When we were tired, we would call out "Good–night–ev–ry–bo–dy!" We would take turns calling out the syllables one by one, and nobody was allowed to talk after that.

In another photo, we are standing proudly beneath the cross that stands at the summit of a mountain in the Austrian Alps, dressed in lederhosen and climbing boots.

My brothers and I became a unit very quickly. I felt very close to them immediately, and I still do today.

Having stayed at home with Inge for the first few weeks after my arrival in Waldtrudering, I soon wanted to join my new brothers at their preschool. I joined the same group as Manuel. In the mornings, the three of us trotted off together, collecting our friends on the way. Even though we were quite young, we usually went on our own. On the way home we always had to prove our courage: We would dare each other to walk closely along a fence, from behind which a big dog—we called him Buddy—would bark at us. My brothers would often send me ahead—I was the bravest of the three.

Waldtrudering is a quiet, middle-class suburb of Munich—a purely residential area where most of the dwellings are one-family homes surrounded by large gardens. The streets are named after German colonies and birds: Togo Street, Cameroon Street, Grouse Way, Birdsong Close. There are hardly any stores or businesses. It caused a stir when a McDonald's opened on the arterial highway that links Waldtrudering to downtown Munich.

For the first few years we lived in a first-floor apartment; then we moved to a single-family home. The rooms were small and full of nooks and crannies. The hallway and stairs were unheated; when you opened a door, ice-cold air would gush in.

In the new house, my brothers and I had a playroom for messy games, but mostly we would play outside in the fresh air. In the summer months, the garden would burst into flower, and a hammock would be strung between two trees. There was a soccer field not far from the house, and a hill. In the winter, we would meet up with other children from the neighborhood and go tobogganing down the hill, tumbling and shrieking with joy. In the evening we would collapse into our beds, hoarse and exhausted.

At the end of the road were fields and meadows, and beyond them lay the woods. We played hide-and-seek there, rode our bikes around, started a club, and built dens in the woods.

My adoptive parents took us children on mushroom-picking courses, where we were taught how to identify the various species of mushrooms found in the woods. For vacations we would go mountain-climbing in Austria or camping in Italy, usually with Inge's parents: our Oma Vienna and her husband.

I saw my mother less and less. In the beginning, she would bring me back to her place every so often, or take me to my grandmother Irene's. I only remember fragments of those meetings, but one occasion is still very fresh in my mind: My mother had collected me from my adoptive parents in Waldtrudering. We were in the car, driving toward Hasenbergl, the district in the north of Munich where my mother lived. We didn't talk much, and I spent most of the journey looking out the window. Then the first apartment buildings came into sight, rows of gray, uniform structures interspersed with public lawns.

When we reached the edge of the neighborhood, my mother parked her car. We got out and walked to her apartment. She darted ahead; I followed, dragging my weekend bag behind me. My mother opened the door, and we were greeted by her barking dog.

Before I had even entered the hall, my mother tossed the dog's leash to me and yelled: "Go take him for a walk!" Anxiously I set off. Outside, I hid from the children who were playing between the clotheslines. I hardly knew them, but they had teased me and called me "pickaninny" on previous occasions.

When I returned with her mutt, my mother dropped onto the sofa and lit a cigarette. She was still angry with me for taking the dog out only grudgingly. I sat down with her and asked: "Hey, Mama, what's up?" "Nothing's up," my mother replied.

❊   ❊   ❊

Before they took Jennifer in, the Siebers had never spoken to Monika Goeth; they only knew about her from Jennifer's CPS file.

After Jennifer came to live with the Siebers, Monika Goeth would call them every now and then and arrange dates to collect Jennifer for a visit, or a visit with her grandmother. The Siebers, in turn, would inform Monika Goeth when anything had happened, such as when Jennifer had her tonsils out. They would give her notice before long family holidays, too.

Now, Jennifer had two "mamas," her foster mother, Inge Sieber, and her biological mother, Monika Goeth.

Inge and Gerhard Sieber had thought about what their foster daughter should call them. Their sons called them "Mama" and "Papa," and soon Jennifer did the same. Inge Sieber would refer to Monika Goeth as "the other mama," as in, "the other mama has got to go to work, that's why you're with us now."

Ruth Irene Goeth visited her granddaughter's foster family once, and the Siebers got on well with her. Monika Goeth, however, would come only to the front door when she picked Jennifer up. Inge Sieber never asked her inside. She found Monika Goeth to be reserved, and wasn't able to warm to her.

Today, Inge Sieber cannot understand why she never talked to Monika

Goeth about her daughter, especially because Jennifer was often restless and troubled after spending a weekend with her mother. "She never told us much about it," Inge Sieber says, "she only ever mentioned her grandmother or her mother's dog."

She recalls that once Monika Goeth did not even come in person to return her four-year-old daughter to her foster family; instead she sent the little girl back in a taxicab on her own.

When Jennifer was six years old, Monika Goeth was expecting a child with her then-husband Hagen. She consented to give Jennifer up for adoption—but not by any family, only by the Siebers.

Since Inge Sieber was not a German national—she is Austrian—the adoption process dragged on for nearly a year. Inge Sieber had to provide a number of references; friends and acquaintances testified that they thought she would do a good job.

For Jennifer and other children of her age, adoption was a complicated and abstract concept. One of Jennifer's little friends told her: "My mommy says that you're bedopted now, no, I mean redopted." And once, when Inge Sieber explained to Jennifer that it would have been physically impossible for her to give birth to Jennifer, because Manuel was only six months older, Jennifer reacted with a child's logic: "Then it's lucky that I was adopted, otherwise I wouldn't even have been born!"

Over the following three years, Monika Goeth kept sending letters and presents to her daughter, but Jennifer's adoptive parents only passed some of them on to her. When Monika Goeth heard nothing back from her daughter, she wrote a letter to the Siebers, asking if it was OK for her to get in touch every now and then, to keep sending letters and presents.

No, they replied—would she please refrain from contacting Jennifer? The girl was too torn between her natural and her adoptive family. It should wait until she was older.

They never heard from Monika Goeth again.

Jennifer Teege and her adoptive brother Matthias on a hiking trip in the mountains

Inge Sieber recalls that it didn't occur to her or her husband to stay in touch with Jennifer's biological mother after the adoption. "We thought that a clean break would be in Jenny's best interest. It was a no-brainer for us: on the day of the adoption, she became *our* daughter."

■ ■ ■

ON PAPER, I WAS A SIEBER NOW. In second grade I wrote a different name on my schoolbooks than I had in first grade. But my mother still belonged to me.

My adoptive parents thought it was best to act as if I really was their own child, as if I had always been theirs.

Yet our story together didn't begin until I was three. I came to them as a Goeth, and they were the Siebers. After the adoption, it was as if my mother had never existed. Suddenly, all contact with her stopped. She no longer called or came to pick me up for the weekend. What had happened? Had she forgotten about me?

My adoptive parents didn't say anything, or encourage me to talk about it. Quite the opposite: Inge and Gerhard seemed glad that I wasn't asking questions.

All they wanted was a normal family.

I didn't dare ask. Was I even allowed to? Wouldn't that mean that I was questioning my new parents? I wanted to belong to the Siebers. When they asked me, at age six, if I wanted to be adopted by them, I said yes.

All I wanted was a normal family, too.

Almost all the photos of my childhood show me laughing: buried in the sand on a beach in Italy, skiing with my brothers, eating ice cream, at the Oktoberfest.

Nevertheless, the smiling photographs don't tell the whole truth.

Early on, I knew: I was different. Different from Inge and Gerhard, my brothers, and the other children. A quick look in the mirror was enough.

Inge and Gerhard talked about me as "our daughter." I know they meant well, but often it was too much. At those words, others would stare at me, mouths agape, obviously asking themselves, "How can that be?" I pretended not to notice their surprised faces.

My childhood photos, the ones I like so much, all show two fair children and a dark one.

In the street, children sometimes called me names such as "Negerbub"—"black boy"—mistaking me for a boy due to my height and short, curly hair. I would quickly retort, "I am a mixed-race girl!" At birthday parties I always hoped nobody would look at me when they were handing out the Mallomars, which were called "Negro's Kisses" in Germany at the time.

At my preschool, I was the only child with dark skin, but in elementary school I met two girls who looked like me: sisters, their father black, their mother white. Just like me. I dreaded the thought that others might lump us together, so I kept my distance on the playground.

Later, in high school, there were two more dark-skinned, adopted children. Maybe I could have discussed my experience with them, but we only talked about everyday things. I had already internalized the silence.

My husband once suggested that we could take in a foster child. I don't know if I'd be able to cope with that. If we did, I would choose a child with darker skin, one who would look more like my own children, and be more likely to feel that they "fit in."

My adoptive parents were idealists. They did not worry about appearances; they just wanted to give a child a second chance. The first family I was introduced to rejected me because of my height; that would have been inconceivable for Inge and Gerhard.

I called Inge and Gerhard "Mama" and "Papa," just like my brothers did. At the time, those words rolled easily off my tongue. Once I became a mother myself, I started calling them "Oma Inge" and "Opa Gerhard." It felt more suitable. They loved being grandparents and were totally taken up by their new role.

After finding the book about my mother, I stopped calling them Mama and Papa. I felt it was important to distinguish between them and my biological parents.

As a young child, I could never say "adoptive parents" without feelings of shame; I never described myself as an "adoptive daughter." The word "adoptive" sounded like a flaw. I was uncertain of exactly what it meant, but I knew it was something awkward. I could look at the adoption certificate whenever I wanted—it was kept with other important documents in the desk—but we never talked about it in the family.

The adoption became a taboo subject.

I didn't even discuss my mother with my brothers, although we had a very close relationship: They were simply my brothers. With them, I could just be myself.

It was easier for them. Unlike my adoptive parents, they did not have to replace my natural parents, to compete with my mother.

Much is expected of foster and adoptive parents—to be and do everything that natural parents would, to become a strange child's mother and father as soon as the child arrives. Yet it takes time to grow into that role. In the beginning sympathy may prevail, as they feel sorry for the vulnerable little creature that has suddenly come to live under their roof. But getting to know the child's personality, and growing together as a family, takes time.

I did not take my adoptive parents' affection for granted. I was afraid I might lose it again.

Inge and Gerhard have always asserted that they love the three of us all the same. But I don't think that's possible. It *is* possible to love every child, but in different ways.

■   ■   ■

The younger of Jennifer Teege's adoptive brothers, Manuel, claims that he never saw Jennifer as his "adoptive sister." "She is my sister. Jenny has been with us for as long as I can remember." According to Matthias, her older brother, her adoption was in fact discussed, "but always in retrospect: That's how it was in the children's home, and then she came to live with us. The question of how Jenny might be feeling about it, or what her mother might be going through, was never addressed."

The subject was avoided, says Matthias, because it would have put a question mark over the siblings' equality. "That was our dogma: Everyone is treated equally. I only realized later that that wasn't the case." In reality, his parents had more trouble with Jennifer than with the boys. "They would argue a lot, partly because Jennifer was a girl. Our mother applied two different yardsticks; she was less tolerant with Jennifer. On the other hand, Jennifer could be undiplomatic. She would provoke our parents or rub them the wrong way."

Inge Sieber noticed it in her own mother: True, Oma Vienna fully accepted Jennifer, but she was always a little more reserved toward her than toward her biological grandchildren, Matthias and Manuel.

Inge Sieber attributes her difficulties with Jennifer to the fact that her new daughter came with a very different personality: "I am more of an anxious type—Jenny is vivacious and confident. I wanted her to come home on time, she insisted on her freedom. We fought so many battles."

■　■　■

ONCE, IN A CANDY STORE, when I was nine or ten years old, I slipped two little marzipan piglets into my pocket. I was caught red-handed by the sales clerk, who told me off in front of all the customers. She made me put the sweets back; my adoptive parents never found out.

A few months later, I pocketed a bag of chocolates in a supermarket. I got through the checkout without being stopped and ran toward the exit—straight into the arms of a large man, the store security guard. He steered me into a side room and made me empty my pockets, whereupon the chocolates came to the surface. He called my parents first, and then the police. I could already see myself sitting in a cell in handcuffs. After a while, Inge arrived.

Looking sad and embarrassed, she talked with the police officers and apologized to the guard. Inge and I drove home in silence. When Gerhard came home from the office, they summoned me to the living room and both gave me a severe dressing-down. I had to promise, hand over heart, that I would never shoplift again.

I went to bed worried sick that they would send me back to Salberg House. Like all abandoned children, I was traumatized by feelings of worthlessness. After all, my original parents hadn't found me loveable enough to keep.

My adoptive parents tried their hardest to be perfect parents, but they could not dispel my fear of being abandoned again. I thought that I had to earn their love over and over. I was missing a basic sense of trust.

One night, I dreamed that my brothers and I were sharing a peach: one half for each of them, which left me with just the stone.

It highlighted my underlying feeling: Whatever my brothers had was beyond my reach.

My adoptive parents set high store by performance and achievement. They taught us the importance of diligence and good grades from an early age. When Matthias was in fourth grade, he took an intelligence test. His results were outstanding; Inge and Gerhard were very proud.

Manuel was in my class. He was also one of the best students and achieved top grades in every subject. My own grades were so-so, and for many years I doubted my intelligence.

I must have been ten or eleven when one day I went hunting through the closets in my parents' bedroom. They weren't home, and I was hoping to discover some hidden Christmas presents.

What I found was a card with a gold chain and pendant. The card was signed, "Lots of love from Monika and little Charlotte."

Little Charlotte—that must be my younger half-sister, I thought, the girl my mother was about to have when she gave me up for adoption.

I did not confront my adoptive parents. I was too embarrassed for having rummaged through their closets.

At least now I knew that my mother was still thinking of me.

At age twelve or thirteen, during an argument with my parents, I demanded to be put in touch with my mother. Furiously, I declared that I wanted to see her again. My adoptive parents explained that I should wait until I was sixteen. Then I would be legally entitled to know my mother's address and to contact her if I so wished.

*　*　*

In the 1970s, it was common practice for adoptive parents to break all contact between their adopted child and the biological parents.

The realization that it is in the best interest of the child's development to deal openly with their past took hold only gradually. But every child has a right to know their origins; it even says so in the 1990 UN Convention on the Rights of the Child.

Today, adoptive parents are advised to explain the reasons for the adoption to their child from an early age, and to keep an album with photos of the child's natural parents, for example. It is also recommended that they find out as much as they can about the child's history. They have to be proactive, since many children don't dare to raise any questions themselves.

Nowadays, adoption clinics are more likely to discuss potential problems a child may encounter as a consequence of adoption. Studies have shown that adopted children are more likely than birth children to feel unloved, have self-doubt, crave recognition, and fear abandonment.

They are often afraid to commit and are more likely to suffer from severe depression and to seek psychiatric treatment.

Often, they will test their adoptive parents sorely: Will they still love me even if I behave really badly? Puberty, in particular, is liable to turn into an endurance test for the adoptive child and parents.

■ ■ ■

GERHARD AND INGE NEVER HEARD ME use the phrase that adopted children typically seem to shout at their parents: "You're not my real parents anyway, you can't tell me what to do!" It wouldn't even have occurred to me, since I was grateful to them. They had taken me in, given me a new life and a future.

But by the time I reached puberty, I was no longer content with just gratitude.

My rebellion against Inge and Gerhard was always partly driven by the question about my mother, by the question of who I really was.

At my adoptive family's dinner table, everybody had their place. Mine was on the left, in front of the windowsill with the flowerpots. But it wasn't only the seats that were set, our roles were as well: Manuel, blond and slim, was always best at everything, the highly intelligent one, while still being friendly and uncomplicated. He was closely followed by Matthias, equally strong at school, calm and bright, but less predictable than his ever-diplomatic brother.

My role was that of the fun-loving goose. When the dinner table conversation revolved around politics or culture, I turned away or yawned pointedly.

Chernobyl, the Cold War—those were the topics of the eighties.

Inge and Gerhard were very interested in politics. Inge was a member of Women for Peace, and the whole family joined anti-rearmament demonstrations. Gerhard, who had always been a loyal follower of the Social Democrats, voted for the Green Party for the first time. They went about saving energy and separating their trash for recycling with ardent zeal (I was the only one who refused to rinse my yogurt cups).

My brother Matthias was elected student body president at our high school. He was much more involved at school than I; he distributed flyers he'd printed himself and painted banners: a Pershing missile with an X through it.

Manuel was very interested in protecting the environment, too; he went on to study geoecology at college. His stickers proclaiming "Nuclear power? No thanks!" graced our bedroom door.

Eventually, we children each had our own bedroom. Mine had a pitched roof with a window opening onto the sky. I pushed my bed right under the window so that I could watch the clouds. I read a lot, spending hours up there with my books; I retreated into my own little world.

My room also served as a kind of "coffee house" where I would meet with my brothers. We would hang a sign on the door reading POD—short for "problem-oriented discussions," and talk about the things we weren't comfortable discussing with Inge and Gerhard: our friendships, broken hearts, fears, and dreams.

When I thought about my mother in those times, I would only recall her positive side; I'd suppress the bad. Lying in my bed at night, I would try to remember what she looked like and picture her long, dark hair. I envisioned her showing up at the door one day, taking me into her arms, stroking me. She would take me home, I fantasized, buy me expensive things, and allow me to do all the things my adoptive parents barred me from doing: wearing makeup or pantyhose, for example, or playing with Barbie dolls.

I soon wanted to get away from my family, from Germany. We weren't a normal family, yet no one seemed to notice but me. At 16 I went on my first vacation without my parents. On an InterRail pass, I crisscrossed Europe with a girlfriend, discovering Paris, Rome, and the Spanish island of Formentera by rail and by boat.

My late teenage years were more carefree than my childhood. I spent less time thinking about my mother and stopped brooding so much. Matthias' and my motto was *carpe diem*—seize the day. I had a large number of friends and went out most nights; I particularly enjoyed parties. On the weekends I worked at a club, the Skyscraper. A bouncer would check you out at the door; if you passed, you took an elevator to the top floor, which offered a sweeping view of Leopold Street, Schwabing's famous boulevard, below. Above the dance floor was a sliding roof; in the summer, people would dance under the open sky. I was 18; I worked behind the bar and thought that everyone there was ultra-cool, me included. I was not allowed to drink while at work, but I could smoke. I smoked like a chimney.

Then one day, when I was 20 and had recently graduated from high school, a girl named Charlotte called our home telephone. Matthias answered the call and took a message for me: "Some girl called Charlotte wants to talk to you." I remembered the card I had found in my parents' closet, and my mother's words: "Lots of love from Monika and little Charlotte."

My half-sister. The girl who had essentially "replaced" me: When my mother became pregnant with Charlotte, she gave me up for adoption. Charlotte came, I went. I did the math in my head: She had to be fourteen now.

I returned her call. A young, friendly voice. Charlotte explained that she would be visiting Munich soon, to see her father. Hagen—the man who beat my mother. He still haunts me in my dreams. Charlotte explained that he and my mother were now divorced.

Charlotte and I arranged to meet in a café the following evening. She had light brown, shoulder-length hair; she was wearing pants and a T-shirt. We talked for hours. She told me about her childhood and asked me, "Is your new family nice?"

She also described how she found out about me. By chance, she had discovered her mother's maternity log, with details of all her mother's pregnancies. Under "Children," she had spotted my name listed above hers. Charlotte ran to our mother and showed her the log. "Who is Jennifer?" she wanted to know. My mother claimed that I had died. Charlotte didn't believe her and probed further. In the end, my mother admitted that I wasn't dead. Only adopted.

My mother told Charlotte my new surname, which is how she found me and my brothers in the phonebook. (We children had our own landline because we would spend so much time on the phone.)

On impulse, Charlotte and I decided to meet again in two days' time. We drove to Starnberg and walked along the lakeshore. The sun was shining, and we spent the whole afternoon together. It was strange having a sister all of a sudden, but somehow it was nice, too. At the same time, I sensed that something wasn't quite right with her.

A few days after my meeting with Charlotte, my mother contacted my adoptive parents. She wanted to meet with me, preferably at our house in Waldtrudering. I didn't like that idea; it was too much—after all these years, I wanted to meet my mother on my own. I suggested that my mother could see just Inge and Gerhard first. I would wait to meet with her afterward, in a café downtown.

■ ■ ■

By the time Monika Goeth saw her daughter Jennifer in 1991, she had been divorced from her husband Hagen for several years. In the end, he had threatened her with a gun, and she had called the police.

She met her second husband, Dieter, at work: Monika worked as a secretary at one of Munich's universities. Dieter was very different from her first husband; he was calm, kind, and friendly. "Like winning the lottery," Monika said. After the wedding, Monika Goeth took on her husband's surname, by which she was known in private and in public from then on.

Dieter was soon offered a job in the country, and the family moved to a small village in Bavaria.

Charlotte started doing drugs at a young age. By her teen years, she was abusing heroin. She spent many years battling her addiction.

Monika Goeth brought Dieter along to the meeting with the Siebers. Inge and Gerhard had set up a table in their garden under the apple tree, and, together with Jennifer's eldest brother Matthias, they welcomed Monika and her husband into their home.

This time, Inge Sieber found Jennifer's mother more approachable. "She was very nice and said that she wished I could have adopted her too, that she would have had a better life then. I was pleased and touched by her praise." Matthias remembers the conversation as rather awkward and stilted. He recalls that Jennifer's mother seemed very nervous.

After a while, Inge Sieber told Monika Goeth: "You'd better go now, our daughter is waiting for you."

I HAD BEEN WAITING IN THE CAFÉ on Wiener Platz in Munich for a while when a woman entered. A man was with her, and I nearly didn't recognize her: She had shoulder-length hair, dyed dark blonde. The last time I had seen her, 15 years ago, she'd had long, dark hair, which I liked better.

She was heading straight toward me. I got up. Uncertain of what to do, we shook hands. I was disappointed that she had brought her husband along. Why hadn't she come on her own? I had hoped for a private conversation.

Nevertheless, I was glad to see her again after all those years. I talked a lot, trying to make a good impression. I told her about my final exams, my recent trip to France, and my plans to visit a friend in Israel. She didn't comment on anything.

She remained guarded and hardly talked about herself. I didn't dare ask her any of the questions that were burning inside me: Why did you give me away? Why did I never see you or my grandmother again? What did Irene die of?

We had lost so much time, and now that we were sitting opposite each other, the distance seemed immense.

We parted with another formal handshake.

I hoped that we would see each other again soon, but my mother didn't call. Not after a week, not after a month, not after a year.

I didn't call her either. It wasn't a conscious decision; I just assumed that she would call me. After all, she was the mother. I didn't call Charlotte, either, since she was living with my mother.

Years after the meeting in the café, I was lying on the couch of the first therapist I saw for my depression. She asked about my mother, and suddenly my head started whirring. Like a film, my last meeting with my mother replayed before my eyes. And I got it at last: My mother didn't want anything to do with me, and she would never get in touch with me again.

Our "good-bye" in the café had not been open-ended. She hadn't forgotten to call me; she just didn't want to. Why not? Did I mean so little to her? For months, I went from grief to rage and back again, but worst of all was the sense of powerlessness.

When I first began to suffer from depression, there were days when I would sit for hours, the photo album from my earliest childhood on my knees, trying to remember everything. I went to Inge and Gerhard's and asked questions—questions I should have asked a long time ago. What was it like when my mother came to collect me? How did we greet each other? And when she dropped me off again, did I hug her before she left? Was there any affection at all? Did they know her violent husband? Inge and Gerhard seemed surprised that I would bring it up now, after all these years. They said that they could not recall any displays of affection between me and my mother, no hugging. They had never met her husband Hagen either.

For many years, I had to deal with those questions on my own. Every now and then I thought about contacting my mother, but I never did.

Now, I have finally written to her, we have spoken. On our phone call, I also asked her about my half-sister Charlotte. My mother gave me her phone number, so I called her and we arranged to meet. Again, I am going to see my sister before I meet with my mother.

I am stunned by how pretty Charlotte is. She is wearing her long hair loose, and she has a beautifully curved mouth. Her appearance belies the difficult years she has behind her. Again we talk about our different childhoods, about our mother. I have the impression that our conversation stirs up memories for Charlotte that she would rather keep blocked out. I witnessed my mother and Hagen's disastrous marriage on my childhood visits to Hasenbergl. Charlotte spent her entire childhood in that broken home; she experienced every row firsthand.

I realize that our conversation is taking a toll on her, and I feel sorry for her. I am enjoying seeing her again, but I am not sure if she feels the same. Am I asking too many questions? Or am I talking too much about my life, about my studies abroad, my travels— all the opportunities she never had? I don't want to hurt my sister.

I am getting angry with my mother again: Why didn't she protect Charlotte? But I have resolved not to be angry anymore. I want to meet my mother openly, without any reservations.

I would love to have a relationship with her. True, we have lived separate lives, but in the end we are still connected. I, too, have carried the burden of our family secret.

■　■　■

Jennifer Teege quickly understood that her grandfather was a criminal. It had taken her mother years to realize that.

Ruth Irene's suicide in 1983 changed Monika Goeth's view of her father, Amon Goeth: "Until then I had always fought against my father. After Irene's death I suddenly felt that I had to protect him—there was nobody else left who would have done that now. I wanted to accept Amon at last so that Irene could be at peace."

In 1994, *Schindler's List* came to German cinemas. Monika Goeth could not watch it to the end. Every time Amon Goeth—portrayed by the actor Ralph Fiennes—produced his pistol, Monika Goeth thought: "Stop it, just stop it!"

After watching the film, she stayed in bed for three days. Her husband called a doctor, who diagnosed a nervous breakdown.

Now Monika Goeth wanted to know all the details: She researched in archives, visited Krakow and Auschwitz again and again. She met with survivors from Płaszów. Monika did not walk to those meetings—she trudged there, burdened by guilt and shame and insecurity. Some survivors told her that they felt anxious near her, that they could not bear her presence since she looked so much like her father.

Regarding her father's deeds, Monika Goeth has said: "I believe all that, but I can't live with it. They hanged my father three times; my mother took her own life. I think that, surely, one day, my life will also come to a violent end."

Monika Goeth underwent a kind of public therapy—but not guided by a psychologist: The documentary filmmaker Matthias Kessler submitted her to a long, torturous interview, confronting her with her father's crimes. He would later turn it into a book with the title *I Have to Love My Father, Don't I?* It was that book that Jennifer Teege discovered in a library in Hamburg in the summer of 2008.

In 2006, filmmaker James Moll recorded the meeting between Monika Goeth and Płaszów survivor Helen Rosenzweig on film. Both women cried when they met in Płaszów. Their meeting is marked by misunderstandings, with Monika still repeating the phrases she grew up with: She explains to Helen Rosenzweig that Amon Goeth only shot Jews because they spread infectious diseases. Helen Rosenzweig is shocked. She interrupts Monika Goeth and asks: "Monika, please stop, stop right now." The film, *Inheritance,* first aired on German television in 2008, the day after Jennifer Teege found the book about her mother.

Monika Goeth later regretted her demeanor in James Moll's film. "I would never try to defend Amon again. I would just be quiet and listen to Helen."

Monika Goeth earned her high school diploma in her mid-forties; later she gained her Latin proficiency certificate and studied ancient Hebrew. She enjoys listening to Israeli music and has read almost all the standard works on the Holocaust. She is nearly 70 now, but she is still fighting the shadows of the past, every day.

※　※　※

IN A COUPLE OF HOURS I am going to meet with my mother.

I am feeling very apprehensive on my way there. I really hope that the curse that has overshadowed our family for so long can be lifted. That we will find peace at last.

My husband and I drive by flocks of sheep in green pastures as we near my mother's small town in the Bavarian countryside. He has come with me, but he won't be joining me for the meeting with my mother; he's going to wait in the hotel room. My mother is coming on her own, too. This time I want to see her alone, just mother and daughter.

We have arranged to meet in the restaurant of the small hotel where my husband and I are staying. I take a seat and wait. At the agreed time, there is no sign of her. At first I don't worry too much and use the extra time to gather my thoughts. But after a while I do begin to feel restless. I go outside to see if she is coming. A little later she arrives at last; she had been stuck in traffic. I am glad that she has made it at all.

This time she seems less like a stranger than when last we met, in the café in Munich when I was 20. I've seen her in the film, so I knew what to expect.

We talk about the little town where she lives, an easy, innocuous subject. Then she looks at me and declares that I remind her of my grandmother Irene—it's the way I dress, my handbag matching my shoes. It sounds like a reproach.

She mentions my grandmother, and, again and again, Amon Goeth—as if it were only recently that he was the commandant of Płaszów, only yesterday that my grandmother committed suicide. My mother says that she is living with the dead.

In an interview, she once admitted that having bad thoughts about Amon Goeth feels like she is betraying her mother, Irene, since he was the love of her mother's life. She believes that she has to be loyal to her mother, and that includes being loyal to Amon Goeth. That leaves her with a terrible dilemma.

Maybe that's the difference between the second and third generations, between my mother and me. My mind is freer than hers. I can have fond memories of my grandmother while also condemning Amon Goeth and my grandmother's life by his side.

I want to shake my mother and shout: "You are living now! Talk to me! Stop talking about your parents. Think about you and me! Don't look back, look ahead!"

I have read many books written by the descendants of Nazis. I understand that my mother's fate is not hers alone, but one that she shares with a whole generation of Nazi children. She is representative of this second generation. Many children of perpetrators have spent their whole lives suffering under their family history. Many have broken homes.

Seeing it like this makes it easier for me: My mother didn't give me away because there was something wrong with me, but because she had her hands full simply dealing with her own life.

The process of coming to terms with the Nazi era has always been a family affair.

The feelings the children of many infamous Nazis have toward their fathers range from glorification on one end of the scale to unbridled hatred on the other, and everything in between. Often, loathing for their fathers turns into self-loathing. The one thing they all have in common: They cannot escape the past.

Gudrun Burwitz, Heinrich Himmler's daughter, was an active neo-Nazi and collected donations for former Nazi criminals. Wolf-Ruediger Hess, son of Hitler's deputy Rudolf Hess, spent his whole life trying to clear his father's name. He proudly announced to his father in jail that his second grandchild had been born on "the Führer's birthday."

Bettina Goering, however, great-niece of Hermann Goering, Hitler's chief of the Luftwaffe, chose to be sterilized so that she would not "create another monster, not produce any more Goerings." Historian Tanja Hetzer says that, in interviews with Nazi descendants, she has learned of other men and women who chose sterilization or childlessness. "In this way, the Nazi ideology of 'worthy and worthless' life is propagated in the second and third generation and is, in an auto-aggressive way, directed toward their own offspring: They don't feel worthy to pass on their genes," Hetzer observes.

Niklas Frank, son of Hans Frank, Hitler's deputy in occupied Poland, carries a picture of a corpse in his wallet: It's a picture of his father with a broken neck, taken after he was hanged for his crimes. Frank says that, every night, he executes his parents afresh in his head, that they deserve it. In his book about his father, *In the Shadow of the Reich*, he writes: "I still feel like my father's puppet, he is still holding the strings in his hands."

Frank claims that his sister Brigitte "died from her father," as if he was an affliction. She committed suicide when she was 46 years old—the same age her father was when he was executed.

Many descendants of Nazi criminals are haunted their whole lives by images of their fathers.

There are many ways to distance oneself from such a father. Karl-Otto Saur Jr., for example, son of Albert Speer's confidant in the Reich's Ministry for Armaments and War Production, used to wear his hair long in defiance of his father's sharp, clean haircut. Monika Goeth studied ancient Hebrew.

■ ■ ■

TODAY, BETTINA GOERING LIVES IN NEW MEXICO. She speaks only in English and has kept her ex-husband's surname. I can understand not wanting to be called Goering, but her decision to be sterilized sends the wrong message. There is no Nazi gene.

I could not finish Niklas Frank's books about his parents, *In the Shadow of the Reich* and *My German Mother*. They may be important documents, but I didn't like them. They are not really about his parents; they are about his suffering from having those parents. Every line in the books is an infuriated cry, full of loathing and self-hatred. Yet all this hatred leads to nothing.

Holding on to the past does nothing to help the victims, nor does it aid in our analysis of—or coming to terms with—our Nazi past.

Ultimately, some children of perpetrators lose themselves behind their overpowering father figures. They define themselves by their past. But who are they when they step out of their fathers' and mothers' shadows? What is left of them? What do they stand for?

Malgorzata, the interpreter who accompanied me to my

grandfather's villa in Krakow, had also shown Niklas Frank and my mother around the house.

Niklas Frank resembles my mother, insofar as his parents became the central theme in his life. My mother is not as outspoken as he is, but I sense that she, too, believes that she has no right to her own life, no right to be happy. My mother believes that she has to atone for my grandfather's deeds, for my grandmother's looking the other way.

Constant self-flagellation and self-damnation will eventually make a person ill—and this kind of suffering, over one's identity and family history, gets passed on to one's children.

I have seen this firsthand in Israel, with Holocaust victims: They buried themselves in their pain and transferred their fears to the next generation. The trauma experienced by the child of a Holocaust survivor is totally different from that experienced by a child of a perpetrator, but the transfer process is similar.

I know that I don't want to live like my mother: arrested in the past, always in the shadow of Amon Goeth.

I think it's a good thing that people like Niklas Frank and my mother go into schools and give lectures about their parents, but that's not the path for me. One day I will tell my Israeli friends and their children my family history. I hope that I will be able to confide in them soon. I want to share my story with them.

I want to walk upright, to live a normal life. There is no such thing as inherited guilt. Everybody has the right to their own life story.

■   ■   ■

The third generation of Nazi descendants usually looks back upon their family with more detachment and fewer false justifications.

The children are still struggling to grasp their fathers' crimes—the grandchildren are trying to come to terms with their family's involvement. They analyze the oft-repeated family legends, investigate what is true and what was contorted or concealed.

The grandfathers' deeds—and especially the silence surrounding those deeds—impacts families to this day. According to the historian Wolfgang Benz, a report on "the history of the mentality of National Socialism and its enduring consequences" is yet to be written.

Historians talk of a "family conspiracy of silence." Katrin Himmler, great-niece of SS chief Heinrich Himmler, refers to the half-truths handed down in her family as "thought prisons." She was able to use her own research to show that Himmler's family directly profited from his position and that some members actively supported his genocidal policies, including her own grandfather, Heinrich Himmler's brother.

Other grandchildren advocate the idea of treating the past as the past. Writer Ferdinand von Schirach, son of Reich Youth Leader Baldur von Schirach, drew a definite line under his grandfather's history in his article for the German news magazine *Der Spiegel*:

"My grandfather's guilt is my grandfather's guilt. Our Federal Supreme Court defines guilt as the result of one person's individual actions. . . . I am more interested in our world today. I write about postwar justice, about German courts who passed cruel judgments, about the judges who sentenced Nazi criminals to five minutes in jail for each murder they committed. . . . We think that we are safe, but the opposite is the case: We can lose our freedom, and then we'd lose everything. It is our life now, our responsibility. . . . 'You are who you are,' that's the only answer I can give people who ask about my grandfather. It took me a long time to find it."

* * *

I DON'T THINK WE DESCENDANTS can disengage entirely from our past; it impacts us whether we like it or not.

I have studied the biographies of many Nazi descendants. The third generation no longer denies what happened under the Third Reich; they speak frankly about it. Nonetheless, some of those stories are lacking something: The people disappear behind the facts, they remain strangers. For me, they are too theoretical in examining their families' past; I find it difficult to identify with them. Dealing with the sins of one's parents and grandparents can destroy families; it is not an academic exercise.

And it doesn't stop here: The past will have an impact on my children, too. My two sons are still young. In a few years, they may well watch *Schindler's List* at school. I don't want them to be ashamed; instead, I hope that they will talk openly about their family history.

I think that we can only get square with our past and truly leave it behind us if we deal with it openly. Feeling that you have to hide yourself and your identity will only make you ill.

That is why I was so shocked when I discovered what my mother had concealed from me: The secrecy that overshadowed her childhood, her youth, her entire life—she let me grow up with it, too. I found out about it much too late.

I'm desperate for her to appreciate my despair, to understand the sadness that I have lived with all these years—to see what a relief it would have been for me to know the history of my natural family.

My mother and I have been talking for over two hours now—and not a word about us. Cautiously, I try to steer her away from the subject of the Holocaust and ask about my childhood.

My mother tells me that she moved back in with my grandmother Irene when she became pregnant with me. The two women discussed the option of my staying with them in their apartment. The children's home was only meant as a temporary measure, my mother explains.

After I was born, she adds, she struggled to bond with me. My grandmother, on the other hand, was immediately taken with me, and she often pointed out what an extremely good grandchild I was, that I never cried or whined. My grandmother loved taking me out on walks or grocery shopping, my mother recalls. Irene, a striking figure herself, enjoyed showing off her black grandchild, like a baby doll. She loved my exotic appearance. Even Lulu, my grandmother's transvestite lodger, would proudly push me around the English Garden in my stroller.

For the first time, I learn the inspiration for my given names: My name is Jennifer Annette Susanne. My mother explains that "Jennifer" was a vestige from the American occupying forces after the war. She liked that it sounded foreign.

"Annette" was my grandmother's suggestion; she just really liked the name.

"Susanne" was in honor of "Susanna" from Płaszów—since Amon Goeth had two maids named Helen, he called Helen Hirsch "Lena" and Helen Rosenzweig "Susanna."

After the war, my grandmother would often talk about her maids in Płaszów. When she was a child, my mother believed that Lena and Susanna were relations of some sort. Only years later did she discover that the years they spent in my grandfather's villa must have been the worst time of their lives.

So, I was named "Susanne" after a Jewish concentration camp

survivor—after the woman who meets with my mother in Krakow in the documentary that I watched so avidly, after the woman whose story moved me so much.

I ask my mother about my adoption. She tells me that it first occurred to her to have the Siebers adopt me when I told her that I wanted to have the same surname as my brothers. She adds that she discussed the matter with my grandmother Irene and that she said, "Sure, why not? I've met the family and they seemed very nice."

For my mother, the adoption was a mere formality. She thought that it would cut the red tape and make things easier for me and my adoptive parents. It was not until afterward that she realized it also meant she had relinquished her visiting rights. She tells me that she was angry and disappointed when she found out.

Once she was not allowed to see me anymore, she would sometimes drive past our house in Waldtrudering, and on occasion my grandmother would come with her.

My mother would look at the big, beautiful home where I now lived and tell herself that it was probably for the best, that she really couldn't ask for more.

She came to terms with the adoption.

My mother only noticed the outward appearances: the house, the garden, the trappings of a middle-class lifestyle. She didn't see how torn I was feeling inside.

She still doesn't see it. She doesn't ask me what life was like with my new family, if there was anything I missed, or if I missed her.

For her, it seems obvious: The adoption was in my best interest. I had a picture-perfect childhood.

She thinks that it was good for me to no longer bear the name

"Goeth," nor the burden of the family history that came with it. She still doesn't see that not knowing was the greater burden.

We talk for nearly four hours. She is learning ancient Hebrew, but she doesn't ask me what it was like living in Israel.

I am trying to put myself in her shoes, to be as careful and understanding as I can, to not ask too much. I turn into the mother, she becomes the child: I feel the need to protect her and help her.

My mother has invited my husband and me to dinner that evening. Her house is in a small hamlet near some woodlands. The garden is immaculately kept; my mother evidently enjoys her gardening.

When my husband and I arrive, my mother and Dieter are still in the kitchen preparing dinner. We share a bottle of wine in the kitchen before sitting down at the dinner table to talk. It is the first time that my husband has met my mother. He, too, notices that her conversation revolves mainly around her parents.

My mother tells us that she once asked Irene: "Why couldn't Oskar Schindler have been my father? Why did it have to be Amon?" And Irene replied: "You wouldn't exist if Oskar had been your father."

Once again, my mother compares me to Irene, but this time it sounds more favorable.

As we are looking together at photographs of my grandmother, she suddenly says: "Pick one for yourself!" I select one that shows Irene in profile; she looks just as I remember her, elegant yet natural, with a scarf draped around her shoulders.

My mother hands me the picture, and then she gives me a small cigar box. I open it up and find my grandmother's favorite golden bangle inside. My mother says: "It's yours." I love the unassuming piece instantly, but I hesitate: I don't want anything from the camp, nothing stolen, no gold from the teeth of the victims. On hearing that the bangle originally belonged to my great-grandmother, I accept it gladly. The gesture makes me happy.

I am accepting of the fact that my mother talks almost exclusively about the past. I think, at the time, that this meeting is just the beginning.

When we arrived, she greeted me with a handshake. As we are leaving, she gives me a brief hug.

I have a mother now.

■　■　■

A few days later, Jennifer Teege is smiling as she talks about her reunion with her mother. She is wearing her grandmother's bangle on her arm.

Jennifer's older brother Matthias says: "After the meeting with her mother, Jenny's natural family was at the forefront of her thinking. She questioned the time she'd spent with her adoptive family."

Inge Sieber experiences this as "Jennifer detaching herself from her family a second time, after puberty: After her reunion with her mother she was very critical of us; that was very hard for my husband and me." It was hard enough for Inge Sieber when, after she discovered the book, Jennifer started calling her by her first name instead of "Mama": "That hurt me deeply." Only sometimes, she adds, does Jennifer forget and call her "Mama" again.

A second meeting with her mother is arranged, this time including her sister Charlotte, too. Once again, Jennifer and her family are staying with her adoptive parents in Waldtrudering.

Her adoptive father, Gerhard, takes Jennifer's sons for a walk in the garden. He has recently planted two trees there especially for them—a gingko for Claudius and an apple tree for Linus. Now he is showing them to his grandsons. "Opa, we're going to meet Mama's mama today," Jennifer's younger son Linus announces.

As Jennifer Teege, her husband, and her sons set off, Jennifer's adoptive parents wave good-bye.

■　■　■

I AM LOOKING FORWARD TO SEEING my mother again. This time, Charlotte will be coming, too, and my husband and our children. A Good Friday out with my family. My family: It sounds authentic when I say it like that.

I wish there was more lightness in these relationships. I hope that we can get beyond the point where all we talk about is our family legacy. It would be nice to just go out together and have fun.

I once said half-jokingly to Charlotte: "Maybe one day we'll spend Christmas together!" I don't think that that will ever happen. I have my adoptive family; they will always be there. But it would be lovely if my mother and Charlotte could become part of my life, too.

My mother has a great opportunity now: After all these years, she can have her daughter back.

It was interesting to talk with my mother on her own; I learned so much about her and my grandmother. I could even fit some of the long-missing pieces into the jigsaw puzzle of my life.

I already knew some of the things she told me from the book about her, but now there is a new character in her story: me. It had hurt me deeply that there was no mention of me in the book. She says that was to protect me, to enable me to have a new life.

She is not the type to question her own actions; she lives too much in her own world for that. She often comes across as aloof and says things that sound harsh and absolute. But I believe that, behind the façade, there lies a woman in need of love, and worthy of love.

I can see the path she has taken: Emotionally, she hit rock bottom over and over again. She had a terrible first marriage—and now she is leading a nearly normal life. I think she keeps many things at a distance deliberately, to protect herself.

My husband parks the car in front of the restaurant where we have arranged to meet. We get out of the car. Charlotte arrives; she looks exhausted. She notices the golden bangle on my arm, and I tell her that our mother gave it to me, that it is from Irene. Charlotte stares at the bangle but says nothing.

My mother and Dieter arrive, and we have lunch together. Afterward, we go for a walk in the English Garden. We hire rowing boats on the lake in the park and later take the children to the playground.

Outwardly it looks like an ordinary get-together, but the atmosphere is quite strained. I would have liked it to be more lighthearted, more personal. I had hoped to get to know Monika as my mother, not just as an older woman. After our first meeting, I still thought that might be possible. Now I am being more realistic; I sense that my mother does not crave a close relationship as much as I do. It's partly due to her personality—she doesn't have a strong maternal side.

I am 40 years old now, she is 64. I am too old for mother-daughter bonding. She cannot give me my bottle or hold my hand while I take my first steps. All those things that she's missed—that we've both missed—we can't make up for them now.

I see my mother one more time, four days after our day out on Good Friday. We meet at my grandmother's grave.

I had asked her to take me to Irene's burial place; I felt a real need to go there with her. We agree to meet at the Viktualienmarkt, a popular market for flowers and delicacies in the center of Munich. I buy some flowers and we head to the Nordfriedhof, Schwabing's cemetery. My mother visits the grave regularly; it seems that she has made peace with her mother now.

We walk through the entrance gate. The cemetery is vast. There are tall, ancient trees, and narrow paths winding their way between the graves. Irene is buried in the same plot as her mother, Agnes Kalder. It is a beautiful grave, very simple. My mother and I plant a few pansies together. This visit means so much to me. In front of Irene's grave, I appeal to my mother: "If you should decide one day that you don't want to see me anymore, I will respect your decision. But please say good-bye and don't just disappear like you did when I was little."

We talk on the phone a few more times after that, and then I don't hear from her anymore. I send her a little package, but it comes back three times with a note: "Refused delivery." When I call her number, no one answers. I let it ring and ring—in vain.

■   ■   ■

The sign on Monika Goeth's front door says SHALOM. It means "peace," but there is no peace here.

Months after her visit to the Nordfriedhof with Jennifer, Monika Goeth is sitting on her patio, under a roof of wild vines. Her husband has baked a cake and made coffee. It is a warm, late summer's day; even the flies have come out again. Monika Goeth is talking nonstop, and every now and then she swats one of the little bugs. Digressing wildly, she tells her story, which revolves around one thing only: her parents, "Amon and

Ruth." Sometimes she is calm and smiling, and other times she furiously rants and raves about her screwed-up family.

Monika Goeth says that she has always tried to protect her daughter Jennifer, to keep her out of that whole "Goeth rubbish."

Monika sees Jennifer, first and foremost, as the daughter of her adoptive parents. Jennifer already has a brilliant mother, she asserts—what does she want from me all of a sudden?

Monika Goeth says that, in her eyes, the people who bring a child up are the parents.

She doesn't understand what this strange daughter wants from her, why she clings to the notion that even a terrible truth is better than silence, that a broken family is better than no roots at all.

Monika Goeth says that she was very pleased to hear from Jennifer initially, but adds that her daughter moves at a different pace. She explains how she felt downright harassed by Jennifer's eager desire to fix everything, to reunite the family. She felt as if Jennifer had written a script for a reconciliation scene that they were supposed to act out with all speed. But it takes time to develop a relationship after so many years, and in the end, she adds, she felt that it was too late.

Jennifer Teege will only ever see her mother again on TV. Monika Goeth has been giving fresh interviews; an Israeli documentary-maker has persuaded her to talk about her role as Amon Goeth's daughter. Time and again, Monika Goeth finds herself in the public eye, even though all she wants, she says, is to be left in peace.

Monika Goeth is now raising her grandchild, her daughter Charlotte's son, who is living with her. Monika says: "He is my life. I do the things for him that I would have loved to do for my father when he was a little boy."

Charlotte chose a traditional Jewish name for her son and combined it with her grandfather's, "Amon"—that is the middle name Jennifer Teege's half-sister gave her son.

I HAVE ALWAYS FELT that I am part of a mobile: everyone connected to everyone else by many invisible strings. If one moves, the others move as well. I am right at the bottom; the main characters are moving around above me.

The central figure is Amon Goeth; he is the source of this evil. Despite being long dead, he makes his presence felt, pulling the strings in the background. My adoption allowed me to escape the family system for a while. I had some peaceful years, but nevertheless I am, and always will be, a part of the whole, an important figure at the end of one string.

I wanted to disentangle the old ties and free our movements today, so we'll be able to adjust to whatever crisis comes next—and all pull through it together.

Maybe I was being too ambitious.

My grandmother's portrait, in its silver frame, takes pride of place on a windowsill in our house, next to photographs of my children and my friends.

Sometimes I visit Irene's grave. I take fresh flowers and light a memorial candle. I don't want to change much here: What would my mother say? After all, she has been caring for the grave for years.

The last time I went, the headstone was overgrown with brambles, but I didn't dare cut them back.

CHAPTER 5

GRANDCHILDREN OF THE VICTIMS:

MY FRIENDS IN ISRAEL

Where are you from—Obama family, eh?

—A JERUSALEM SPICE TRADER TALKING TO
JENNIFER TEEGE IN 2011

AM BACK IN ISRAEL. AT LAST.

Tel Aviv has grown: The highways seem twice as wide; new high-rise buildings loom against the sky. Many buildings in the city center remain dilapidated, their façades dirty, corroded from pollution and the salty air. Next to them, the recently renovated buildings gleam all the more brightly.

On Rehov Engel, a small, quiet side street, the house where I shared an apartment in my early twenties lies nestled between palm trees and flowering shrubs. This is where I sat in front of the TV all those years ago and watched *Schindler's List.*

The sun, the salty air, the throaty Hebrew sounds—it's all very familiar. But I am a different person now.

When I arrived here over twenty years ago to visit my friend Noa, I was young, curious, and unburdened. Now I am returning as the granddaughter of Amon Goeth.

Noa was the reason I came to Israel in the first place. She is also the reason I haven't dared to come back here in the three years since I discovered my family's past.

I first met Noa in Paris. After my graduation from high school in 1990, I moved to the French capital for a year. I minded the children of a French family, and I attended classes at the Sorbonne and at an art college, where I was developing a portfolio. I planned to study graphic or communication design on my return from Paris.

I met Noa in a life drawing class. We were both trying very hard to get the model's proportions right. After class ended, we stood in the corridor and talked for a long time.

I liked her quick wit and her sense of humor. She had long, curly blonde hair and light green eyes. She talked openly about herself and her feelings. Noa told me she sometimes had days that were somehow different—she called them her "camera days." I knew what she meant; I also have days where I feel like I am walking through life as a silent observer, zooming in and out on the world. Noa was able to give a name to the feeling that I had always found so hard to put into words. I liked her precise language and her unique view of the world.

Noa was in her early twenties, like me, and she was in Paris to accompany her father, an artist who had been granted a scholarship there. Noa's father always wore black clothes in winter and white only in the summer.

Noa's mother was a lawyer. She had worked in Germany, even in Munich. As a teenager, Noa had tagged along with her mother a number of times. Now she was trying to remember the few German words she had picked up at the time: "Bitte. Danke. Guten Tag."

She wrote down the Hebrew alphabet on a paper napkin for me. I was surprised to learn that words were written from right to left. Noa talked about Israel as if it were a perfectly normal country.

After my year in Paris, I returned home. I hadn't been accepted at any of the various colleges where I had applied to study design. I decided to visit Noa in Tel Aviv. When we had said our good-byes in Paris, she had asked me: "When are you coming to Israel?"

It was a four-hour flight from Munich to Tel Aviv. Noa was waiting for me in her apartment. She beamed at me when she opened the door and briefly showed me her room, which I was going to share with her. Her roommate, Anat, was sitting on the balcony. She was a little older than Noa, with strawberry blonde hair. "Anat, this is Jenny," said Noa by way of introduction. Anat and I shook hands. Soon Noa was eager to go out: She wanted to celebrate our reunion somewhere special.

Noa hailed a taxicab, and we drove along the promenade toward the south of Tel Aviv. After half an hour, the cab stopped along a gravel path. I got out and looked around: We were standing on a cliff top, the sea below us. In front of us was an open-air bar, full to bursting.

The bar, Turquoise, had deck chairs and swing seats on the lawn. More and more people were streaming inside. I thought the women were stunningly beautiful; almost all of them had long, dark, curly hair.

In hindsight, it is difficult to say what I had been expecting, but it certainly wasn't this: carefree people enjoying themselves, sitting in swing seats surrounded by palm trees and brightly colored beach umbrellas, listening to chill-out music, and looking out to sea.

Primed by my history lessons and news reporting about Israel, I had been prepared for a permanent state of emergency. I was thinking Holocaust and intifada (the Palestinian rebellion against the Israeli occupation). I had expected a traumatized people—in a country where bombs could explode anywhere and anytime.

I found a city built on sand, its bright apartment blocks on stilts in order to allow the fresh sea breeze to fan the city streets.

I had expected heaviness and found lightness instead.

That afternoon I knew: *I want to stay here, in this country.*

Noa and I sat down on the warm grass. We watched the sun as it set slowly over the sea.

■　■　■

Tel Aviv means "Hill of Spring." It got its name in 1909, when Jewish immigrants founded a small settlement in the sand dunes by the Mediterranean Sea.

In the years following 1933, many European Jews sought refuge in Palestine. They wanted a country of their own, because in so many other countries they feared for their lives.

The first wave was of refugees who were able to flee Germany and the territories occupied by German troops in time.

Others did not escape from the Nazis. They came to Palestine after the war, mentally and physically broken by their experiences in camps like Płaszów.

The exhibition rooms at Yad Vashem, Jerusalem's Holocaust memorial, are deliberately kept in semidarkness. Visitors are presented with evidence and testimonials of Nazi crimes, among them a photograph of concentration camp commandant Amon Goeth, sitting astride his white horse in his SS uniform. There are more pictures of other commandants and other concentration camps—at last, visitors grasp the extent of the killings. And then the corridor leads them slightly uphill to the exit, where they emerge into a sun-drenched Israel. This is Yad Vashem's message: Israel and the Holocaust are like day and night. The new country is the emergence into light.

On May 14, 1948, Zionist leader David Ben-Gurion proclaimed the establishment of the state of Israel in Tel Aviv. That day, the Jews danced with joy on Dizengoff Street, Tel Aviv's main boulevard.

While the establishment of the state of Israel meant that the Zionists' dream of a Jewish state had come true, the Zionist myth of "a land without a people for a people without a land" was at odds with reality: The Jews' holy land was already inhabited by Palestinian Arabs. Two peoples tied their national identity to the same piece of land. What for the Israelis is the founding of their country—*Yom Ha'atzmaut*, "Independence Day"— is, for the Palestinians, *Nakba*—"The Catastrophe."

That same night, troops from five Arab countries—Egypt, Iraq, Transjordan, Lebanon, and Syria—marched into Israel with the proclaimed aim of destroying the newly declared state. Israel emerged as the winner of this first Arab-Israeli war, and with a considerably enlarged territory. The new borders included a much greater area than Israel had originally been granted by the United Nations. Around 700,000 Palestinians fled or were driven away; many Palestinian villages were destroyed.

This first war immediately following the founding of the state would be but one of many military disputes between Israel and its Arab neighbors.

Israel's history is inextricably tied to the history of the Palestinians; it is a succession of wars and terrorist attacks, of violence and counterviolence. It is impossible to recount it objectively, let alone in a few lines.

One reason why the Middle East conflict seems so intractable is that it is an entanglement of many different conflicts. It isn't just about the land. The Israeli historian Tom Segev writes: "The conflict is fueled not only by political, strategic, and economic interests, but also by fear and jealousy, faith and prejudice, myths and illusions."

Religious fanaticism, stoked by Muslim fundamentalists and ultra-Orthodox Jews, has long pervaded any dialogue on both sides.

Israel, the small country on the edge of the Mediterranean Sea with a current population of eight million people, cannot find peace: Externally, Israel is struggling with its Arab neighbors; internally, with itself. There is no peace between Israelis and Palestinians—but there is no peace between the deeply religious and the secular Jews either.

The deeply religious Orthodox Jews and modern, liberal Israelis are engaged in a bitter struggle for the prerogative of interpreting many political issues: What role should religion play in politics? How to deal with the Palestinians? Should the ultra-Orthodox settlers withdraw from the occupied territories, or should they remain and continue expanding their settlements under the constant guard of young Israeli snipers?

When Jennifer Teege arrived in Tel Aviv in late 1991, the city was already regarded as a haven for modern, democratic Israelis.

Tel Aviv has always been the center of the country's creative scene: Artists and writers come to live here; record labels, advertising agencies, and computer firms are headquartered here. Lefties and liberals, gays and lesbians love the city for its unorthodox way of life.

In the years after the creation of the state, so many Jews came to Tel Aviv that most of its dwellings were raised with lightning speed. Previously, young European architects had designed and constructed around 4,000 Bauhaus-style buildings here and thus created the "White City." Today, Tel Aviv is home to around 400,000 people, originating from over 100 different countries.

Tel Aviv is a place for new beginnings, a city without memories. It is a counter-world to the capital, Jerusalem, where every stone tells a story and religious fanaticism is rife.

Tel Aviv is young and modern, noisy and hectic, open-minded and tolerant. But it is not beautiful and picturesque like Jerusalem. In the summer, the stench of exhaust fumes and trash hangs over the city, and scrawny cats forage for food on the beach.

Israelis say that the most important word for living in the city is *lizrom*—be spontaneous, go with the flow. Jerusalem is the place for praying, they say, Haifa for working, and Tel Aviv for living.

Especially in the nineties, the city cultivated its image as the cool and buoyant party capital. It was often referred to as "the Bubble" thanks to the carefree attitude that prevailed on Tel Aviv's creative scene—despite the fact that the first intifada had begun in 1987, the same year that Palestinians in Gaza established the radical Islamist organization Hamas. Bombs were already exploding around the nation on a regular basis.

In early 1991, during the Gulf War, the Iraqi dictator Saddam Hussein repeatedly fired rockets at Tel Aviv and Haifa, although Israel was not officially involved in the conflict. Jennifer's friend Noa recalls: "During the Iraqi attacks we lived in constant fear for our lives. The sirens wailed again and again, announcing new attacks. For over two months I had plastic sheets taped around my windows to make them airtight. We had gas masks at the ready and enough stockpiled mineral water to last for weeks."

The Iraqi attacks ended almost a year before Jennifer Teege arrived in Tel Aviv. Noa says: "We had been through terrible times, so we were enjoying life even more. It was a carefree time, the most playful time I remember." At Tel Aviv's parties, suffering and danger seemed but a distant notion.

■　■　■

THE NEXT DAY I WAS UP EARLY. So far I had seen Tel Aviv only through the windows of a taxicab; now I wanted to explore the city on foot. I went to a café and ordered what I believed to be a typical Israeli breakfast: freshly squeezed orange juice and a bagel.

After breakfast I walked to the Tayelet, the seaside promenade,

In Israel, 1992

where tall hotel towers line the seafront. Noa, who couldn't come with me because she had a lecture to attend, had suggested a must-do: "Get a view of the city from above!"

I stole into a four-star hotel on the beachfront and took the elevator to the top floor. The view was spectacular: On one side I could see Tel Aviv stretched out in front of me, from the skyscrapers in the center all the way to the suburbs. On the other side was the beach and then the sea.

I went down to the sea. Beachgoers with wet hair were walking around; joggers were running along the water's edge. Israeli pop music was playing in the beach cafés. Children were building sandcastles; surfers were paddling in the waves. I took my shoes off and dropped onto the hot sand.

Later I joined the crowds at Carmel Market, where stall-holders were hawking their wares: fruit and vegetables, underwear, fake Rolex watches. Opposite the market lies Shenkin Street, which, according to Noa, was the trendiest street in the Middle East: cafés and boutiques, records and designer clothing. The way back to Noa's apartment took me via Rothchild Boulevard, an impressive avenue where people of all ages gathered to play *boules* and talk about politics.

I wanted to explore every corner of this country and get to know its people. I was particularly looking forward to seeing Jerusalem: the golden cupola of the Dome of the Rock, the gleaming silver one of the Al-Aqsa Mosque. So far, all I knew about the city I had learned from books and stories; now I was going to see it with my own eyes. At Tel Aviv's Central Bus Station I boarded a *sherut*—a shared taxi—and headed for Jerusalem.

Through the dirty minibus windows I was barely able to make out the blurred landscape: bare hills interspersed with little villages, and every so often soldiers and military roadblocks. Tel Aviv is only 45 miles north of the Gaza Strip, and 35 miles west of the West Bank. I realized how small Israel is—much smaller than I had expected.

An hour later, we reached Jerusalem.

The bus let us off just a few meters outside the Damascus Gate at the northern side of the old city wall. I stopped in front of the massive archway. Armed soldiers with machine guns, helmets, and bulletproof vests were patrolling the battlements. When the soldiers noticed me looking at them, they fixed their gaze on me. I quickly turned away and hurried through the gate into the Old

City's Arab quarter: tight little alleyways, narrow townhouses, arcade shops. The air was scented with tea and herbs. Shopkeepers approached me, trying to entice me in. Children in blue school uniforms capered between the shops, and small traders pushing handcarts laden with fresh fruit and vegetables had to constantly stop and brake wherever the narrow alleyways became too steep.

I didn't see any Orthodox Jews in the Arab quarter of the Old City. Here, the Muslims generally lived among themselves. In the vicinity of the Wailing Wall, further east, it was a completely different picture: Devout Jews were rushing through the streets. They were wearing black trousers and long overcoats, their prayer tassels trailing below.

At the Wailing Wall, I was surprised to see people walking backward. Later I learned that they consider it disrespectful to turn their backs on the Temple and only turn around after they've backed away a few meters.

Men and women pray separately at the Wall. The men had donned their black-and-white prayer shawls and were reciting their prayers while swaying forward and back. Next to them were the women, at a smaller area of the Wall reserved for them. Their lips were moving silently in prayer.

I joined the women and touched the smooth, worn limestone with my hands. It is a Jewish tradition to insert messages and prayers into the cracks between the stones. Without thinking, I wrote a quick wish and squeezed it in with all the others between two stones of the Wall. Slowly I backed away. I had written that I wished for a boyfriend.

■ ■ ■

Jerusalem, the Holy City. A city sacred to three world religions and there-fore fiercely contested like no other.

Here, Muslim, Christian, and Jewish pilgrimage destinations lie side by side: After Mecca and Medina, the Al-Aqsa Mosque and the Dome of the Rock are the Muslims' most important shrines; Christians pilgrimage to the Church of the Holy Sepulchre, where according to tradition Jesus was crucified and buried; and the Wailing Wall, the last remaining wall of the Second Temple, is the most sacred site recognized by the Jewish faith.

To whom does Jerusalem belong? The question has been one of the key issues since the beginning of the Middle East conflict. Jerusalem's legal status is unclear. From a practical point of view, the city is divided into a predominantly Arab eastern part and a Jewish western part.

But it is also divided between the deeply religious and the secular. Almost half of all strictly Orthodox Israelis—known as *Haredim*—live in Jerusalem. Their quarter, Mea Shearim, is reminiscent of a nineteenth century shtetl. The residents converse in Yiddish, a dialect based on me-dieval Middle High German. German speakers can understand it to an extent.

Haredi Jews believe in the strict separation of men and women in pub-lic. On buses, for example, women have to travel at the back. Many also oppose the secular state of Israel; they study the Torah and don't work. Around sixty percent of Haredi families in Israel are living in poverty.

Strictly Orthodox Jews often have very large families. Secular Israelis worry about the fact that in many Jerusalem schools, Orthodox Jewish children are in the majority. The influence Orthodox Jews have on Israeli politics is steadily growing.

■ ■ ■

MEA SHEARIM MEANS "City of a Hundred Gates." Noa and Anat had told me about this quarter of the religious Jews. They said that secular Israelis would avoid the area, and that on Shabbat, the weekly day of rest when observant Jews refrain from work activities, Mea Shearim residents would sometimes throw stones at strange cars approaching the area.

I wanted to see how the people there lived. I walked past derelict buildings and labyrinths of courtyards. Laundry strung across balconies was flapping in the wind.

The Old City of Jerusalem had a certain museum-like quality about it, an air of days gone by, but at the same time it was colorful and vibrant. Mea Shearim, however, felt dark and forbidding. The dwellings were built closely together. There were many children here, too, but they averted their eyes when they saw me looking at them.

The men wore hats or fur caps on their heads, their sidelocks curling down underneath.

The women wore ankle-length skirts and sandals with dark stockings. At first I wondered why all the women had the same pageboy haircut; then I remembered what Noa had told me: Orthodox women usually marry very young; subsequently they shave their heads and wear wigs in public.

At every other crossroads in Mea Shearim there were signs in various languages warning tourists not to enter the area with bare arms or legs.

The people in Mea Shearim live without radio, TV, or Internet. Instead, posters are plastered to the walls of the narrow townhouses, announcing the local news to the residents: the opening of new shops, public lectures, weddings. And some unusual messages: The rabbis were inviting the community to pray for rain, next to a warning that boys and girls are not allowed to walk through the district together.

How could they live like that, in a world full of commands and prohibitions, thrown back to the nineteenth century?

Noa had told me about a friend of hers who married an Orthodox man. Noa was very sad about it: "She lives on a different planet now."

I took a sherut back to Tel Aviv. It was late when I arrived. There were only a few people in the streets. It was Friday night—Shabbat had begun. Candles were burning in the windows of many apartments; families were setting the dinner table together.

Noa and Anat didn't celebrate Shabbat. We sat talking together for a long time that night. I wanted to know more about their country. Anat suggested I join a kibbutz for a while and work alongside volunteers from all over the world for free board and lodging. During her military service, she had lived on Kibbutz Eilot in the south of Israel, where she met her boyfriend, Alon. The next morning, Anat called the kibbutz and arranged a place for me.

■ ■ ■

Anat's partner, Alon, grew up on Eilot. A few weeks after his birth in 1965, his parents took him to the children's house at the kibbutz. There, he lived with the other children, cared for by nurses and teachers. He saw his mother every afternoon before returning to the children's house for bedtime. He still remembers how much he missed his mother at night.

In the early years of many kibbutzim, the patriarchal nuclear family—father, mother, children—seemed like an outdated model. Instead, the task of raising the children was delegated to nurses and teachers. Women were expected to work, just like the men. Housework was centralized, too: The kibbutzim had communal laundries, tailor shops, kitchens, and dining halls where everyone ate together.

The central idea behind the kibbutzim is collective living. The founders of the first kibbutzim were inspired by socialist and Zionist ideas. They wanted to create a socialist Jewish state on their own soil.

The first kibbutz was founded on the shores of the Sea of Galilee, over a hundred years ago. In 2014, there were around 270 of these villages all over Israel.

Kibbutz Eilot was founded in 1962. It lies at the southern tip of Israel, on the edge of the Negev desert, surrounded by rugged mountain ranges, between Jordan and Egypt.

Israeli writer Amos Oz spent 25 years living on a kibbutz. In sum, he says: "The founders hoped to change more than just the social system, the class society. They wanted to revolutionize human nature. They believed that if they created a society where everyone ate the same, dressed the same, worked the same, and shared the same living standard, all selfishness would disappear and a new human being would arise. That turned out not to be the case."

When Alon finished school, he did not receive any vocational training or academic education. He took it for granted that he would work in the workshops on the kibbutz.

■　■　■

I caught a bus that set off for the south. After three hours' journey the driver stopped, in the middle of the desert. All around me there was nothing but red dust and rocks. I marched up the road and found the kibbutz on a hill. At first glance it looked like a holiday complex: identical, two-story buildings separated by green lawns, flagstone paths, and rampant oleander bushes.

The workday began at six a.m. I had hoped to help milk the cows; I thought that might be more exciting than, say, sorting tomatoes on a conveyor belt. It was not to be. On the first morning, I was sent to the kitchen. My job for the day was to dispose of leftovers and to pre-rinse the dirty dishes. On day two, I was moved to the dishwasher. Loading vast numbers of dirty plates into the machine was incredibly tedious, and I wondered if this was really the best way to get a more in-depth picture of Israel.

The next day, I packed my bags and left.

I traveled a few miles further to Israel's southernmost city, the beach resort of Eilat by the Red Sea. Looking to boost my travel funds, I headed to the port, an alleged insider's tip amongst penniless backpackers. A large sailing yacht was just putting out to sea. I ran along the pier and shouted to one of the crew: "Are you looking for some help?" "Sorry, we are full!" he shouted back.

Then I saw a wide, red boat enter the harbor, with a tall, dark, and lean man at the helm. When the last passenger had disembarked, I approached the captain: "Hi, do you need some help?" He smiled: "Yes, someone left this morning."

His name was Shimon, and I fell in love instantly. He had a weather-beaten face with bright blue eyes under heavy eyebrows. Shimon was a Sabra—the term for native Israeli Jews, after the Hebrew name for the prickly pear: hard and spiky on the outside, but soft and sweet on the inside.

Shimon was forty-eight and married with a young daughter. He had spent half his life serving in the Israeli Defense Force, including in Shayetet 13, the elite unit of the Israeli Navy. He had moved to Eilat a few years earlier, and now made his living by operating a glass-bottomed boat between Eilat and the Egyptian border.

I became a member of the crew, together with two other back-packers, a Dutch girl and a South African man. During the day, I would sell tickets to tourists; in the evenings I'd scrub the deck and the toilets. At night, I would fetch my camping mat, unroll my sleeping bag, and find a place to sleep on the boat, under the open sky.

Shimon and I came from different worlds, but we had things in common: We both enjoyed our own company, and we loved tranquility. Shimon took me on a trip to the Negev desert. I had never been to the desert before, but I loved it straight away. It may appear bleak at first—a vast, barren emptiness. But there is so much to discover in the barrenness. We walked through narrow gorges. Shimon showed me unusual rock formations and pointed out how the color of the rock changes over the course of the day. We discovered plants in the desert, and saw snakes and scorpions.

At first it didn't bother me that Shimon didn't talk much. I could have walked through the desert with him in silence for days.

For the first time, someone loved me for who I was.

After a few weeks, however, my usual restlessness kicked in. Did our relationship have a future, or was I wasting my time in Eilat? Shimon didn't understand what I was worried about. True, he was still living with his wife, he said, but the marriage was long finished. He asked: "Why don't you just stay here?" He suggested we move in together in Eilat.

I asked for some time to consider and flew back to Germany. My tourist visa was only valid for three months and about to expire. Back home in Munich, my adoptive parents didn't push me or ask when I might start college. They must have known that I didn't feel like talking. My friends, on the other hand, spoke quite bluntly: They asked me what I, at twenty-one, wanted with such an old man, and suggested I was looking for a father figure.

I did some temp work with Siemens to earn some money. Then I flew back to Eilat.

Shimon's wife had moved out of the apartment, leaving their daughter with us. She was four years old, a beautiful little girl with long, dark curls that had a tendency to fall over her eyes. In the mornings, she went to preschool; in the afternoons, she was either with me or with her mother.

Shimon worked on the boat. I wished for more independence; I wanted to earn my own money. I had a job waiting for me at the holiday club Méditerranée in Eilat, but the season had not yet started. In the meantime I was supposed to be learning Hebrew, but after five minutes I would usually put my books to one side and stroll over to the beach or to the nearby shopping mall.

In the evenings, when Shimon came home, his daughter would leap up and fling her arms around his neck. She came first. He would play with her and read her long stories before putting her to bed. Afterward he would collapse, exhausted, on the sofa. I felt abandoned. This was not how I had imagined my life with Shimon.

One evening, after I had been sitting on the sofa waiting for him yet again, it all came bursting out. I assailed him with grievances, complaining that his life revolved around his daughter and adding that I had cheated on him with an old boyfriend during my recent stay in Germany.

Shimon looked at me calmly. He said that his behavior wasn't the problem, nor was his daughter. "You don't know what you want. Why are you wrecking everything?" Then he leaned back. An oppressive silence hung in the air.

Sometimes I think that if I had met Shimon at a later point in my life, we might have had a future. But at the time, I was too young. I was looking for a savior, not a partner. It was asking too much.

The next morning, I packed my bags and went back to Tel Aviv. I had called Noa in the night after my argument with Shimon.

Shimon didn't contact me. After five days, I went to a travel agent and booked a flight to Germany for the next morning. That night, I set the alarm for 4:30 a.m.

Warm sunlight was falling on my face as I woke up. I squinted. I could see the packed bag by my bed; I could hear banging in the kitchen. I got up and stuck my head out the door. Noa was at the stove and gave me a cheery wave. After we had breakfast together, I went to a language school for newly arrived immigrants and signed up for a Hebrew class.

■　■　■

Noa remembers how Jennifer arrived crying, with her bags at her feet, at her door in Tel Aviv: "She was so in love with Shimon, and so desperate."

When Jennifer's alarm went off on the day of her planned departure, Noa tried to get her out of bed, "but Jenny just mumbled that she was tired and rolled over."

It still makes Noa laugh to think about that morning: "She overslept once—and stayed for four years!" Jennifer was vivacious and spontaneous, Noa says about her friend. "Jenny used to make up her own Hebrew words—we laughed so much."

Jennifer was a great friend, both for having fun with and for meaningful conversations: "When I first met her in Paris, we felt very close very quickly. Our friendship is one of the most extraordinary friendships I've ever had, full of wonderful times and crazy coincidences. I always tell her it was a godsend that she missed that flight."

■　■　■

Hebrew is a Semitic language. Unlike with English or French, it was impossible for me to deduce the meanings of new words by myself. My teacher at the Ulpan, the language school for newly arrived immigrants, made every effort. She was able to explain new words using gestures and facial expressions. If someone didn't understand the expression for "to lie down," she lay down on the desk. It didn't take long for me to be able to follow simple conversations. But it was a long time before I had the courage to speak in Hebrew. Time and again I became exasperated by this new language and its complicated grammar.

I found my own place to live. I met Tzahi, an actor, and we rented a three-bedroom apartment on Engel Street together. At the time, Tzahi was not yet the successful actor he was to become, but he was very popular with women. He was in his mid-thirties, blond, bright, and handsome. Many people thought that we were together, but Tzahi was like a brother to me. We often cooked together or played "guess the capital" while we were doing the dishes. Our roommates changed often, but he and I remained the solid core.

Once I had finished my language course, I applied to study at Tel Aviv University for a degree in Middle Eastern studies and African studies. When the acceptance letter landed in my mailbox, it took a huge weight off my mind. Up to that point, my future had been uncertain. Now I knew—I was going to college in Israel!

In the lectures I sat among Israelis. The lecturers spoke Hebrew, so at first I hardly understood anything at all and spent a lot of time reading up on the topics. I was allowed to take the exams in English. For my Middle Eastern classes I learned Arabic and translated parts of the Koran. Often I would stay up until midnight, my head bent over a book at my poorly lit desk.

I had a new boyfriend, Elias. He had been sitting behind me in one of my Arabic classes and kept staring at me. When we talked during the break, we hit it off straight away. I soon gave him the key to my apartment. I tried to forget about Shimon, but I couldn't.

In the little spare time I had outside my studies, I often met with Noa and Anat. I had become close friends with Anat, too. She was just there, in her quiet, caring way. I once went on a trip to the Sinai Peninsula where I drank tea with the Bedouins in the desert and contracted a bad infection. Back in Israel, I was hospitalized for several days. When I was discharged, I was still weak and feverish. Anat would come and sit by my bedside—and she spent hours in the kitchen making me chicken soup.

I admire Anat for her modest, unpretentious way of living. She eventually moved in with her boyfriend Alon on the kibbutz near Eilat, where I had only stuck it out for a couple of days. Today, Anat works as a nurse; I can't imagine a better profession for her.

■　　■　　■

Anat describes how she spent many hours with Jennifer, walking through Tel Aviv: "We would talk—about Israeli politics or Israeli men. I usually put on my platform shoes because I'm so short and Jennifer is so tall. We stood out when we went out together." Anat also recalls that Jennifer would often be approached by strangers when they walked on the beach: "Some people wanted to know whether Jennifer was a professional basketball player. Or model scouts would come up to her and want to book her for some photo shoot or other. But she always replied, 'No, I am studying.' She was never one of those naïve, clueless girls. She was very independent and seemed to know what she wanted. Friends would often come to her with their problems and ask her for advice."

■   ■   ■

I SPENT A LOT OF MY TIME at the Goethe Institute in Tel Aviv. It was before the days when German newspapers were available on the Internet. I borrowed piles of books, and read literature about the Holocaust, Zionism, and the Middle East conflict. Soon I was known by everyone at the institute, and ultimately I was offered a part-time job in the library. From then on I would go to work there in the mornings and to college in the afternoons.

The Goethe Institute mostly attracted young Israelis who came there to learn German. But older Israelis came, too: Holocaust survivors who wanted to read and hear the German language again. They didn't talk about what they had gone through, but I saw the tattooed numbers on their arms. Initially, I was quite self-conscious and felt that I had to apologize for being German.

The color of my skin was good camouflage: Most visitors to the institute assumed that I was American or maybe one of the many Ethiopian Jews who were arriving in Israel at the time. But once I opened my mouth and started speaking in fluent German, they knew. When I mentioned that I was German to my fellow students at college, they looked at me in amazement: How could

I be from Germany? they wanted to know. How did I end up there?

Some of the Holocaust survivors who came to the Goethe Institute had failing eyesight. If I had the time, I would sit down with them and read German newspapers and novels aloud.

Much later, when I discovered my family history, I was glad that I had read to the old people then. I hadn't done it because I felt guilty, but just because I wanted to. At the time, I had no idea that my grandfather had murdered people just like them.

Two elderly ladies came regularly. They asked about my studies, and we talked about everyday issues. I didn't dare ask them about their history. I just spoke with them in German and told them about today's Germany—so different from the one that they had known.

Once, at an event held at the Goethe Institute, I was talking with a sixteen-year-old Israeli boy who told me that he had been learning German at school for the past three years. I praised his pronunciation. And then, out of the blue, he said that almost his entire family had died in concentration camps. I was ashamed and didn't know what to say, but eventually I blurted out: "Well, what really counts is that you and I can be here together now and talk."

I think he noticed my embarrassment. In any case, he smiled at me and asked if I had ever been to Berlin, or if I had heard of the German punk rock band Die Toten Hosen.

The abrupt change of topic seemed bizarre, but at the same time it was characteristic of that generation of young Israelis. To them, Germany was both the Nazis and the present. They asked me about Boris Becker, Helmut Kohl, and reunification. It hadn't been long since the Berlin Wall came down.

Nathan Durst is a psychologist and deputy chairman of AMCHA, the Israeli center providing assistance and counseling to Holocaust survivors and their families. He sees differences between second- and third-generation descendants of victims of the Holocaust.

Most of the children would have grown up with their parents' silence on the subject, he says: "Some of them would have been named after a murdered relative; yet many parents would not talk to their children about their experiences—because they didn't want to relive the horrors, but also because they were ashamed of the humiliation that they had suffered." But the children could still sense that their parents had gone through terrible ordeals, he adds, speculating that this may be why the second generation still struggles to entertain the idea of reconciliation. "The children of survivors often did not want to have anything to do with Germans. They felt hatred and wanted revenge."

According to Durst, the third generation regards Germans in a different light, clearly differentiating between the past and the present. "Often it is their grandchildren that the victims open up to about their time in the concentration camps, about the experiences that they had locked up inside for so long. It has a healing effect on the whole family: Once things are out in the open and being talked about, they can usually be dealt with much more easily."

EACH YEAR IN THE SPRING, Israel observes Holocaust Remembrance Day. Sirens wail all over the country; for two minutes everything stops, and people remember the victims of the Holocaust. I always found these two minutes very moving, the silent reflection very powerful. I would stand there, among the Israelis, and feel that I was part of the Jewish community of mourners.

With my Israeli friends of my own age, my nationality and the past were irrelevant: Noa and I talked about everyday life, about college. Our friendship thrived on frivolity and the typical things twenty-year-olds care about, such as betting on which of his many female admirers my handsome roommate Tzahi would take to bed next.

During the early days of my stay in Israel, I devoted myself to learning everything there was to know about the Holocaust. Later, the here and now came to the fore. While I was in Israel, apartheid was abolished in South Africa. Since I was also taking African studies at college, this was an important subject for me.

In Middle Eastern studies, we looked into the different conflicts in the Middle East. Unlike my fellow students, I often traveled to the occupied territories, to the West Bank and the Gaza Strip. Hearing about the Palestinians' lives from my Israeli lecturers wasn't enough for me; I wanted to talk with the people living there in order to gauge their situation realistically.

The first time I went to Gaza, I went with a friend who worked for a Palestinian aid organization. I remember how shocked I was when I saw the squalid and derelict homes and streets in Gaza. Posters with PLO Chairman Yasser Arafat's face where everywhere.

The situation in the refugee camps was unbearable. People who had fled their homes decades ago were still living in temporary accommodations. Children were gamboling around the tents, but I didn't see any playgrounds or parks, only dust and desperation. The Palestinians I spoke with kept saying, "Life and death are in Allah's hands." I didn't ask who was to blame, but I couldn't forget the suffering of the people there.

During the four years that I lived in Israel, the political situation escalated. One morning I was waiting at my bus stop in Tel Aviv, part of my daily commute to the Goethe Institute. A line 5 bus arrived; the waiting passengers got on. Shortly afterward, I boarded a different bus. When I reached the Goethe Institute twenty minutes later, I knew straight away that something bad must have happened. The institute is right next to the Ichilov Hospital, and ambulances were speeding past me, sirens wailing and blue lights flashing. Inside the institute, my colleagues were already standing in front of the TV, watching the breaking news. A suicide bomber had blown up the line 5 bus in the center of Tel Aviv. The pictures showed pools of blood and the obliterated bus.

I had taken that bus on numerous occasions. I suddenly realized that I could lose my life in this country that wasn't my own.

■　■　■

The suicide attack on the line 5 bus on Dizengoff Street in October 1994 marked the arrival of political violence in Tel Aviv. The bombing shook the city that represented a modern Israel. Twenty-two people died; forty-eight were injured.

It was the first major attack carried out by the radical Islamist group Hamas since the signing of the Oslo Accords in 1993 and 1994, when Israeli Prime Minister Yitzhak Rabin and Yasser Arafat, chairman of the PLO, had agreed on the withdrawal of the Israeli Defense Forces from parts of the West Bank and the Gaza Strip. The agreements provided for the creation of Palestinian self-government and the renunciation of violence. However, particularly contentious issues, such as the question of the Jewish settlements in Israeli-occupied territories, the status of Jerusalem, and the return of Palestinian refugees, were deliberately kept out of the agreement to be decided at a later stage.

In 1994, Rabin, Arafat, and Israel's then-minister of defense, Shimon Peres, were awarded the Nobel Peace Prize. In a later speech, Yitzhak Rabin promised to "end, once and for all, a hundred years of bloodshed."

Rabin was severely criticized for his policies by the radical Israeli right, while the Islamist organizations Hamas and Palestine Islamic Jihad attempted to disrupt the peace process with terrorist attacks.

On November 4, 1995, Yitzhak Rabin was assassinated at a peace rally by Yigal Amir, an Israeli right-wing extremist.

The bombing of that packed Israeli bus was not the last; many more were to follow. From then on, no one in Tel Aviv or anywhere in Israel could board a bus, or visit a café, bar, or shopping mall, without fearing an attack.

When Yitzak Rabin died, hope for peace in the Middle East died with him.

■ ■ ■

I HADN'T ANTICIPATED THE ATTACK on Yitzhak Rabin. The country was in mourning. An Israeli had killed another Israeli. The danger hadn't come from the outside—not from Gaza, the West Bank, or Lebanon—but from within Israel's own ranks. Rabin's assassination revealed how disjointed the country was.

My view of Israel had been changing for some time: My initial euphoria had been replaced with a deep skepticism. I was living in a highly armed country, surrounded by hostile neighbors. I was aware now of the serious threat that the country was facing, the intractable conflict it was engaged in. And of how one's view of the world became very one-sided for those that lived here.

I can't say exactly when the depressions began. All I know is that I found myself walking through Tel Aviv on my own. No longer happy and outgoing, but sad and introverted. I felt no joy and no curiosity. It was as if a wall had appeared between me and my surroundings.

When I breathed in, I didn't get enough air. I felt like I was being choked.

I grew more and more withdrawn; I wanted to be on my own. I only left the house if I absolutely had to, to go to work or to the library for my studies. I didn't discuss my worries with Noa or Anat; I would not have been able to explain the state I was in.

What was wrong with me? I could find no obvious reason for my sadness. I wasn't homesick; I received regular visits from my friends and my adoptive family. My degree program had been the right choice, too—at last I was doing something that I was really interested in.

No matter how hard I thought about it, there was no explanation for my unhappy state of mind. I scolded myself for being ungrateful. I had seen how people lived in the Palestinian refugee camps. I, on the other hand, was living the good life; I had all I could possibly need. Why couldn't I appreciate that? Why did I find everything so hard?

Perhaps, I thought, the reasons for my sadness could not be found in the present. Perhaps they dated back to something in the past.

I barely managed to concentrate on my final exams. The harder I crammed, the less I seemed to remember.

. . .

Jennifer's adoptive family came to visit her during that time. Her brother Matthias was shocked when he saw her: "Jenny was completely exhausted. I was alarmed by her lifestyle: She had thrown herself fully into her studies, obsessing about details with an exaggerated intensity—as if she wanted to prove something to us." Matthias had the impression that Jennifer wanted to demonstrate to her adoptive family what she was now capable of: "As if her worthiness depended on her being the best."

Noa and Anat also noticed that Jennifer wasn't well, but she didn't want any help. Noa says: "We were really close friends, but Jenny still liked to sort out her problems by herself."

. . .

SOMEHOW I MANAGED TO PASS my final exams. Afterward, I invited Noa, Anat, and a couple of other friends over for dinner. The following morning, I left Israel.

I went back to Munich and started therapy. I took a part-time editorial job with Bayerische Fernsehen, the Bavarian television network. Shortly after my twenty-seventh birthday, I suffered a nervous breakdown—during a conversation with my boss, I started to cry and couldn't stop.

I spent the following days in bed, the duvet pulled over my head. When a friend of mine called, I picked up the receiver and told her to call me back in six months. I didn't want to see anyone, only to stay in bed and sleep.

People who have never suffered from depression cannot imagine what it's like to be depressed. They may assume that depression is like an ordinary emotional low: For a while you "don't feel so good," but at some point you'll start to feel better.

I didn't start to feel better. I fell into a deep hole. My breathlessness became more and more frightening; I was gasping for air and thought I was going to die. When I was at my worst, I would have preferred to die. I never seriously considered taking my own life, but I hoped that I might cross a road and get run over by a car, so that everything would be over.

I had applied for a postgraduate course at the London School of Economics and been offered a place, but there was no way that I could accept it in my current condition. Instead of going to university in London, I went to therapy three times a week.

The first kind of therapy I tried was classic psychoanalysis. I lay on the couch and talked about what was going on in my life or what was on my mind at the time. Often I would talk about things from my past or about my dreams.

During the many hours I spent with my Munich therapist, the subject of my mother resurfaced. My underlying feelings of being given away and abandoned had not been resolved, but merely suppressed. I also suddenly discovered an interest in who my father was.

At elementary school, the other children had always wanted to know where I was from and why my skin was so dark. I would tell them that my father was an African chief who rode through the jungle on elephant-back. Later, I would claim that my father was Idi Amin, the cruel dictator who controlled Uganda in the seventies. He was the only African ruler I knew of as a child. I thought it would make the other children understand that I didn't want to talk about the subject and that they would leave me in peace.

When I moved to Paris after my high school graduation, I spent days wandering through the streets of Goutte d'Or, the African neighborhood in Paris's 18th arrondissement. At the market, stall-holders sold sweet potatoes and cassava roots next to smoked pikes that looked like shriveled rubber. Street vendors offered roasted peanuts and corn on the cob. The women, dressed in brightly patterned, tie-dyed wraps, carried their children on their backs and their shopping on their heads. At the hairdresser's, women would have their long hair braided. One of the market stalls flew a Togolese flag.

Africa suddenly seemed very near.

It was a strange world to me, but at the same time I had a sense of homecoming. I liked the beat of the African music and the kaleidoscope of colors. Here, finally, people didn't look back at me over their shoulders or stare at me from the corner of their eyes.

In Germany, black people are a minority. When we run into each other on the street, we nod and say "hello" even if we don't know one another. Our skin color creates an affinity.

In the African quarter of Paris, the color of my skin was nothing out of the ordinary. For the first time in my life, I felt that I was among my own kind.

I inherited my dark skin from my father. *Where was he now?* I wondered. Who was he? And who was I?

I decided to find my father, and so I contacted CPS. He was living in a village in Germany.

I sent him a little note to find out if he wanted to see me. A few days later I received his reply, on lime-green stationery, in ornate handwriting and polished German. He thanked me for my letter and said that he had always hoped, and even expected, to hear from me one day. Now a huge weight had been lifted from his mind. He would be excited to meet me, he added, and was looking forward to getting to know me at last, to catching up on everything he had missed out on over all these years.

We arranged to meet in a restaurant. When he arrived, he presented me with a rose.

My father is Nigerian. He told me that he was from Umutu, a small town in the southeast of Nigeria, and that he belonged to the Igbo people, an ethnic group in Nigeria. Originally forest farmers, today they are predominantly traders, craftsmen, and civil servants. The majority of the Igbo belong to the Christian faith.

My father told me that, when he set off for Germany in the late sixties, he was one of the first people to leave his village. At the time, anyone in Nigeria who aspired to a career sought a Western education. Besides, the country was ravaged by civil war.

After he studied in Germany, my father returned to Nigeria, where he worked for the government. He explained how the corruption drove him to despair: Computers designated for schools ended up with employees of the ministry. Eventually he moved back to Germany. Today, he is married to a German woman and has five other children. My half-siblings.

When I was born, he wanted me to be brought up by my African grandmother in Umutu. I learn that my father gave me an African name, Isioma, a traditional Igbo name meaning "Lucky."

My father gave me two books by the Nigerian writer Chinua Achebe, which I enjoyed very much. The books are about African traditions and about a kind of personal god that the Igbo people believe in: the *chi* that determines one's life. When a person loses their way, their chi will try to lead him back to the right path, Achebe explains.

I, too, wonder if life is just a succession of coincidences, or if we are guided by some higher power like chi. For a long time I didn't believe in fate, only in chance. But since I learned about my family history, I have thought differently. We are not free in our decisions— some things about our journey through life are predetermined.

After our meal at the restaurant, we went our separate ways: My father returned to his family, and I went back to Munich, to my old life.

As a child, I had known my mother, and so I missed her. My father, on the other hand, had always been a stranger to me. I had been curious about him and wanted to meet him in order to understand more about myself. But I had never felt a longing for him. Our meeting didn't change my feelings for him, either. He remained a stranger to me.

I saw him once more, when he invited me to his home. I met his family, his wife and children. I could see that my father was making every effort, but I was overwhelmed by all these new people. We said a cordial good-bye. I didn't see him again for a long time.

A few months later I moved to Hamburg. A friend had told me about a new media agency there. I wanted to get away from the heaviness of the political issues at the television network. I thought a job in advertising would be more lighthearted. By then, I felt emotionally strong enough to hold down a regular job.

■ ■ ■

The photo Jennifer Teege included with her application for the job in Hamburg shows her wearing a summer top and massive sunglasses. She also sent in a number of ideas for TV and magazine ads, as well as her report card from first grade: "Jennifer has integrated well into the class."

Her application suited the agency, and it suited the times. In the late nineties, the new economy was still booming. The agency was hiring new people every month; there was plenty to do. Hairdressers and masseurs would come to the office; in the mornings there was breakfast for all. The workday started at 9 a.m., and if you went home at 6 p.m., your colleagues in the open-plan office would ask: "Just doing a half-day today?"

■　■　■

ON MY FIRST DAY AT THE NEW JOB, a tall man with a deep voice approached me in the corridor: "Are you new here?" He was the agency boss. Goetz. The man I would go on to share my life with.

Newly in love, I would sit in a top-floor office in the center of Hamburg, writing copy for online marketing campaigns for banks and tobacco, car and furniture brands.

I enjoyed my work. The atmosphere was good, everybody was in high spirits. The campaigns for such a diverse range of products gave me the perfect excuse to act out my curiosity at last. I have always quizzed the people around me; I'm interested in how other people live—from where they go on vacation to what sort of bed they sleep in, what kind of sofa they sit on.

But it didn't take long for the same old problems to raise their heads again. It turned out that I still didn't have a grip on my depression. It was no longer a constant issue, but it came in bouts. At some point, every new task would cause me to panic. I would spend days playing around with old texts in order to look busy. Once I pretended I had the flu to avoid having to go to work.

Telling the truth was not an option. The world of advertising is all about perfect façades. No one talks about mental health problems, since they are counterproductive to creativity. At the time, I felt under constant pressure. Once a week, I went to a psychological self-help group. On those evenings, I always skulked off early under some pretext.

■　■　■

Goetz Teege is a quiet, levelheaded man of few words. Regarding his wife, Jennifer, he says: "It was love at first sight. She inspires me." Goetz Teege comes from a stable family and has four siblings. He tries to relate to the kind of upbringing Jennifer had: "Her fundamental problem is that she has learned not to rely on anyone." Despite her difficult childhood, his wife has "enormous strength," he says. "I was always fascinated by that side of her: Whenever she was feeling low, she fought hard to get over it. She sought to understand, to get to the bottom of it. She always felt that there was something that she didn't know about. Something that she needed to find out in order to get a handle on her life."

■　■　■

MY RELATIONSHIP WITH GOETZ BECAME STEADY. Soon, we were talking about children. He was seven years older than I and already had two children from a previous marriage; he wasn't sure if he wanted to become a father again. If he had decided that he didn't want any more children, I would have ended our relationship. I could not imagine a life without children. At age 32, I gave birth to my first son; two years later I had my second.

I endeavored to give my sons all those things that I went without for years: warmth, security. Normality.

Today, the most important thing I want them to come away

with is a strong sense of self-worth. I don't want them to have to work as hard for it as I did, in hundreds of hours of therapy.

At first, I found it incredibly hard to leave the children with anyone. I didn't have the heart to say good-bye to them. Whenever the babysitter came, I would sneak out to spare them the pain of separation.

Today, I would do things differently. I have come to understand that children can tolerate a brief good-bye, a short separation. It is much worse to leave without saying good-bye at all: If children find their mother suddenly gone, it erodes their basic sense of trust.

■ ■ ■

Jennifer Teege's brother, Matthias, notices that his sister is a very tense and anxious kind of mother to her sons: "She is very protective of her children, maybe too protective."

Matthias believes that Jennifer demands too much of herself: "In Israel, she strove to be the perfect student. Now she aims to be the perfect mother." Jennifer's perception of the model mother is a mother who is there for her children 24/7, he explains. "She tries to offer them the kind of childhood she never had. She tries to be the mother she would have liked to have."

After her marriage to Goetz and the birth of her sons, Jennifer Teege's depression gave way to a more bearable sadness. Her life seemed secure—until the day, at age 38, she found the book about her mother.

Suddenly, Anat and Noa stopped hearing from their friend. Anat says: "We would normally be in regular contact, but then it just ended abruptly. For months, we didn't hear a word from Jennifer. Noa and I were really worried; we kept sending her emails: 'What's up with you? Please write to us!'"

■　■　■

AFTER I FOUND THE BOOK about my mother, I couldn't bring myself to write to Noa and Anat. I needed time to recover from the shock.

When I was ready to turn to my friends in Israel, I realized how hard it was. It felt as if I had led some kind of double life all these years. As if I had been lying to my friends and all the people around me.

Even though the family secret was not my fault, I had a guilty conscience. I was particularly scared of telling Noa; I didn't know how she would cope. Some things affected her deeply.

Had Noa lost any relatives during the Holocaust? We had discussed the subject during my studies in Israel. I knew that none of her close family members had been murdered, but I knew nothing of her more distant relations. Had anyone been killed at Płaszów? If she had mentioned it at the time, I might not have taken note.

I would have felt more comfortable telling my story to Anat; she is not unsettled so easily. But I needed to talk to Noa first.

And so I didn't confide in either of my friends, and I only answered their emails sporadically. We didn't see each other for nearly three years.

Every year at Rosh Hashanah, the Jewish New Year, Noa sent me photos of her family. Sometimes she would write to me on the occasion of other Jewish festivals, too, or for family events. I usually replied with just a few short sentences.

Finally, Noa announced that she was coming to the next Berlinale, Berlin's annual international film festival. Noa had become a screenwriter in Israel. Every time she came to the Berlinale, we would try to meet up; our reunions there had become a little ritual. I hadn't been in touch for a long time—if I didn't go to Berlin now, she would think that I was deliberately trying to avoid her.

But on the other hand, I couldn't go to the festival and talk with Noa about trivial matters. I couldn't and wouldn't lie to her if she asked what was up with me. We had known each other for too long for that.

The Berlinale was going to show a film for which Noa had written the screenplay; it was about an autistic boy. My old roommate Tzahi played one of the main characters.

I knew how long Noa had been working on this screenplay—for years. She was always talking about it. Now, she had invited me to Berlin; she wanted me to sit next to her when it was shown at the cinema. It would be her big moment—one I wanted to share with her, not destroy with my story.

I had once made the mistake of telling a good friend my family's story at a birthday party. My friend had become so upset that she couldn't enjoy the rest of the party.

I wrote a long email to Noa's husband Yoel and explained the difficult situation I found myself in: There was something on my mind that I needed to discuss with Noa, but I didn't want to tell her at the film festival. I wrote down the whole story and asked him to share it with Noa. I also asked him how many relatives he and Noa had lost in the Holocaust, and if anyone they knew had died in Płaszów.

Yoel wrote back: "We have all lost someone. The Holocaust is in our DNA, it is why we are here. But how is that your fault? The

Berlin film festival is a great moment for Noa, but you won't spoil it for her. She is longing to see you again, she's missing you. I am sure that she will listen to your story and help you however she can. You don't need to spare her. You need us to support and look after you now, not the other way round. Noa will always be your friend, in good times and bad."

■ ■ ■

Yoel and Noa are sitting in their apartment in the center of Tel Aviv as they tell their family stories.

Noa's father's family was living in the US when Hitler came to power; they were safe.

Noa's mother's family was from Poland and Russia. Her maternal grandmother was living in Stolin, Belarus, when the war broke out. She was deported to Siberia under Stalin. Her parents, her four siblings, and their children stayed behind and were killed by Germans in a massacre, along with hundreds of other Jews.

Relatives of Noa's maternal grandfather were killed in the ghetto of Pinsk in what was then Poland. His brother died in the Majdanek concentration camp, near Lublin.

Noa's husband, Yoel, also lost relatives in Poland. He recalls how in the seventies, when they were still children, he and his friends were surprised to learn that a neighbor owned a Volkswagen Beetle. This neighbor had been imprisoned in a concentration camp—and now he was driving a German car!

That was a long time ago, Yoel says. He laughs and points to his stove: Siemens!

In Yoel's hometown, there were couples who adopted children because they could not have children themselves: They had become infertile due to the abuse and medical experiments they had suffered in concentration camps. They were severely traumatized people, living in constant fear that their adoptive children would one day disappear or be taken away from them.

Yoel very carefully told his wife what he had learned from Jennifer's email. Noa says that she was shocked: "I had never been so intimate with a close relative of a Nazi criminal." Noa had other German friends, too, and wondered what crimes their grandfathers might have committed during the war.

Why had she never asked? "At first I didn't dare ask about their grandparents. And later the subject was so far removed. If you are friends with someone, you don't discuss whether their grandparents may have killed or informed on your grandparents. It's particularly striking in Jenny's case: Her and Amon Goeth—I just can't wrap my head around it!"

Noa is convinced: "It was fate that I met Jenny as a young woman. It would have been impossible to strike up a friendship if we had known that her grandfather was a concentration camp commandant. How could she have come up to me, with such a rucksack full of guilt? How could I have met her without any bias?"

It would have been so complicated, she adds—a tense, "reaching hands across the graves" sort of friendship.

Today, Noa explains, she can deal with Jennifer's story. She has known Jennifer for twenty years and sees her only as a friend, not as the granddaughter of a Nazi: "I told her, forget Amon Goeth. You are Jenny! Please, come!"

■　■　■

I AM BACK IN ISRAEL NOW, at Noa's. She has moved; I had to hunt for her new apartment. After we embrace and Noa shows me

Jennifer Teege and her friend Noa in a café in Tel Aviv

around, we sit in the sun on her patio, talking and watching the goings-on in the street below. It is like always, only better, since now there is nothing between us.

A few weeks ago I was sitting next to Noa in the dark, watching clips of her film at the Berlin film festival. It was just as I had hoped—sharing that significant moment in her life.

Now, in Tel Aviv, Noa's film is showing at the Dizengoff Center downtown. That evening, we go to see the full-length movie. It is called *Mabul (The Flood)*. We see Tzahi on screen, playing the father. I like the story; it's about a family with an autistic son. It shows how important it is to stick together and not give up, even in trying times.

After the movie, Noa and I go to a café. We talk about all the things we have done and been through together. We are closer than ever. There is nothing left to hide; everything feels good and right.

After a quick visit to Jerusalem, I travel to Eilat. To Anat. I had asked Noa to tell her everything.

◼ ◼ ◼

Anat cried when she heard Jennifer's story.

This scene from *Schindler's List* instantly came to her mind: a man on his balcony, shooting people as a pastime. Jennifer's grandfather.

Anat switched the film off at the balcony scene. She couldn't bear it.

When Anat shows old, faded family photographs, she often explains, "he was shot," or "she was gassed."

Anat's mother's family was originally from Poland. Anat's great-grandparents and an uncle were probably killed in Sobibor—a death camp in the then-district of Lublin where Amon Goeth was posted temporarily before he came to Krakow.

Anat's father was a German Jew from Hanover; he escaped to Israel in 1935. His relatives who stayed behind in Germany were all killed.

After the war, Anat's father returned to Germany only once. When he came back to Israel, he told his children, "They are still the same, they haven't changed."

Anat's father hated the Germans, and he hated God for allowing it all to happen. Anat grew up with a bitter, old man for a father. But shortly before his death, he would suddenly only watch German TV and only listen to German radio stations.

Jennifer and Anat are sitting side by side on the porch of Anat's house in Kibbutz Eilot. Anat has made fresh mint tea and put dates on the table. She is barefoot, as are most people here; her blonde, shoulder-length hair is disheveled, and she is wearing a baggy T-shirt.

Children are running across the well-tended lawns. Nowadays, young families choose to live on a kibbutz because they want their children to grow up in a safe environment, close to nature and alongside other children of the same age. The communal nurseries are still here, but the children live at home with their parents. Anat says that she would not have

Jennifer Teege with her friend Anat and Anat's second son, Stav, in Kibbutz Eilot in 2011

joined Alon on the kibbutz if their children would have had to grow up in a children's house, as her husband did.

Today, Kibbutz Eilot is evocative of a modern townhouse complex anywhere: Children are laughing and cats are meowing; everyone knows everyone. But it has retained its strong communal spirit: There are no hedges or fences between the individual homes, and everyone pays their income into the communal pot. At the end of the day, there's not much left for the individual.

Jennifer Teege took the desert road to come here, past Bedouin settlements and signs warning of camels in the road. The longer she drove through the Negev desert, the more relaxed she seemed to become.

It has been a long journey to get to this point. She has put Krakow behind her, and a small village in Bavaria.

Jennifer and Anat are holding hands; Jennifer is stroking Anat's hand. Anat has put on Jennifer's enormous sunglasses—a fashionable model that doesn't seem quite right for her. "I'll be the talk of the kibbutz in these," Anat says and laughs.

Anat's elder son, Kai, is seventeen now—nearly the same age Anat and Jennifer were when they first met. For the last two years, Kai's history lessons have been mainly about the Holocaust. Anat says that her son is now filled with rage against the Germans.

Soon, Kai will go on a school trip to visit various concentration camps in Poland—standard practice for Israeli teenagers. Anat would like Jennifer to go with them, would like a German with her particular history to be with Kai's class when they visit Płaszów.

■ ▩ ■

I HAVE TO THINK about whether I want to accompany Kai and his classmates on their trip. I'd like to look forward now, not back.

We walk through the kibbutz. Anat shows me the new guesthouses. The next time I come, I want to bring my husband and my sons. I've always wanted to visit Israel as a family—but not until my children are old enough to understand this complicated country.

I hug her as we say good-bye: "Anati, my dear friend."

CHAPTER 6

FLOWERS IN KRAKOW

Everybody wants to know who they are.

—JENNIFER TEEGE'S FORMER THERAPIST AT THE UNIVERSITY MEDICAL CENTER HAMBURG–EPPENDORF

WHAT IS FAMILY? Is it something we inherit, or something we build?

It's been exactly four years since I discovered the book about my mother, three years since I visited Krakow for the first time. When I first traveled to Poland, I had hit rock bottom. Reading the book about my mother had reopened old wounds: the hurt from my childhood, the feeling of not knowing who I am, and the sadness that had overshadowed my whole life.

Everybody wants to know where they come from, who their parents and grandparents are. Everybody wants to be able to tell their complete story, with a beginning and an end. Everybody asks: *What is unique about me?*

The book was the key to everything, the key to my life. It revealed my family secret, but the truth that lay before me was terrifying.

I went to Krakow to get closer to the overwhelming figure of Amon Goeth, to understand why he destroyed my family.

During my last visit, three years ago, I didn't have the courage to admit my identity to a Jewish tourist I happened to meet. I couldn't even tell my friends in Israel who I really was.

Those days are behind me now. I have returned to Krakow to meet my friend Anat and her son Kai. Anat has come with the class to Poland, along with a few other parents and the children's teachers.

Tomorrow, I will stand in front of Kai's class and tell them my story. How will the children react?

At first, I wasn't at all sure whether I really wanted to join them. I didn't speak to Anat about my concerns; my doubts had nothing to do with her. It was just that I had resolved not to talk about the Nazi era all the time. Not because I think it's the wrong thing to do—I think it's good that the descendants of Nazi criminals urge people to question how they are dealing with the past. But I don't want my life to revolve around just the subject of Nazism. There is an unending number of issues that are worthy of support, and I am not an expert on the Holocaust.

I decided to fulfill Anat's request anyway. After all, I wouldn't be addressing any class; it is her son Kai's class, and I'll have Anat by my side.

And I thought that it might be interesting for them to meet me. I didn't consider the impact the meeting might have on me. I had no expectations whatsoever and just wanted to take things as they came.

When I arrive at Krakow airport, I am exhausted. I haven't had time to properly prepare for my meeting with the Israeli students. I had planned on reviewing some vocab so that I could give my introduction tomorrow in Hebrew. That is out of the question now.

I have just come from the deathbed of my adoptive father. He died a few hours ago in the St. John of God Hospital in Munich. The cancer started in his prostate, but by the end it had spread all over his body.

I hail a taxicab and head toward the city. It is getting dark. In my head, I am still going over the last few days at the hospital, still sitting by Gerhard's bedside. In the past two weeks I have come to realize what it means to die. Before then, death was only an abstract concept for me.

Never before had I been with someone during the last days and weeks of their life.

It only takes a few days for a person to take their leave of this world. The body deteriorates gradually, step by step. There are so many little markers on the way to death. It is a process, and by the end every last thing has been taken from us.

When a close relative dies, we think about our own lives, too. Our own mortality, which we prefer to ignore, is suddenly very real.

When Gerhard was admitted to the hospital, he was still able to feed himself and to briefly lift himself up from the wheelchair. At first he could still drink independently, then only using a straw, and in the end not even that. He was put on a drip and a ventilator. He didn't want any life-extending procedures.

In the beginning, I would get ice cream for him; asking him which flavor he would like. "Strawberry," he would reply, or "mango" or "lemon." He could hardly eat, but the ice cream was his little pleasure.

On the day before he died, he could hardly talk anymore. When I asked him what flavor ice cream he wanted, he couldn't decide. I got some lemon ice cream for him and carefully fed him small spoonfuls. His mouth was so dry, his face so gaunt—he looked like death itself.

Gerhard kept wanting to sit up in bed because it was easier to for him breathe in that position. But the doctors said that we were not allowed to sit him up: The risk of his suffering a collapse was too great. And so he would lie there, with his pleading eyes, begging us to sit him up. Seated opposite him, I felt so helpless. For him, sitting up would have meant that last bit of independence. I would have loved to do him that small favor, but even that was not allowed. At some point he gave up and lay still, with his eyes closed.

At the hospital, all the people who were close to Gerhard came together. Someone was always with him: Inge kept watch at night; during the day, my brothers, friends, and relatives would join her, as well as my husband, our sons, and I.

The only thing that was real was Gerhard's dying; everything else seemed far away. Just like the terminally ill patient who was slipping into timelessness, into the twilight of the in-between world, we, his companions, also lost all sense of day and night.

Gerhard had enough time to say good-bye to his friends and family, properly, consciously. A blessing.

The big question was whether or not Gerhard would make it to his seventieth birthday. That had been his last wish—to celebrate his birthday surrounded by his family.

On the morning of his birthday, we all gathered around his hospital bed. My brother Manuel's daughter had baked a cake.

Gerhard opened his eyes only briefly. He was barely conscious, but he sensed that we had all come. I think he still enjoyed his birthday; it was a nice way to say good-bye. We spent the whole day taking turns by his bedside. Quietly I was hoping that he would die soon. I knew that's what he wanted, too.

Shortly after I left the hospital, Gerhard passed away.

I was onboard a plane three hours later.

Three years ago, I hadn't let a recent miscarriage prevent me from traveling to Krakow. And now it hadn't occurred to me to cancel on Anat.

My journey to Poland had been a long time in planning. I had booked my flight to Krakow weeks ago. It is part of my nature to honor my commitments.

The taxicab stops in front of a large hotel in the Podgorze area of Krakow, the former Jewish ghetto. Tomorrow, I will meet Anat and Kai, Kai's class, and his teachers. It has been nearly a year since I last saw Anat and Kai in Israel. I am looking forward to seeing them again.

■   ■   ■

Płaszów is the penultimate stop for the Israeli students on their tour of Poland.

In the last few days, they have visited the former Warsaw ghetto and the former extermination camp Treblinka, northeast of Warsaw. They have cleared Jewish graves of dirt and leaves and have talked at length with Auschwitz survivor Zvi Moldovan, an old, friendly Israeli who has been accompanying students on school trips to Poland for years.

At the defunct train station in Lodz, the students boarded an old cattle car which was used to transport Jews and Romany from the Lodz ghetto to the concentration camps. Inside the cattle car, it was dark and crowded. The teens tried to relate to how the Jews locked in this car must have felt. One of the Israeli girls began to tremble violently.

The students visited the Chelmno concentration camp, and later Lublin and the Majdanek camp.

They visited the former Tarnow ghetto and the Belzec camp.

The class shared a joint diary, in which every student was invited to write. One student wrote: "Numerous members of my family were murdered there. When I looked at the gas chambers, the barracks, and the crematoria, it seemed as if they had only been built yesterday. But I went there in an air-conditioned bus; my grandfather in a hot and crowded cattle car without food or water. My journey took a couple of hours; Grandfather's took three days and three nights. I was there with many of my friends; Grandfather was all alone. I left after a few hours; Grandfather left in the summer of 1944. In Poland, I saw my grandfather's memories with my own eyes. He survived and told me all about his experiences. I will always remember him."

Another student commented: "My grandmother never really escaped the camp. She was always restless and lived in constant fear of losing control over her life again. She planned and prepared things a long time in advance, forever filling her fridge and her larder. I never saw her sitting still."

Halfway through the trip, the teachers gave the students letters that their parents had written in advance. The mothers and fathers had written comforting words, asking their children not to let the terrifying sights send them into despair. Many students started crying when they read their parents' letters.

The last entry in the class diary reads: "Every day here feels like a week. I miss my parents and home so much."

At Sobibor, the students walked through the woods where the escaped prisoners once went into hiding after the unsuccessful uprising in the camp.

They went to the village of Markova in southeastern Poland, where a Polish peasant and his family once hid two Jewish families on their farm. The Germans eventually discovered the Jews and shot them dead, along with their Polish hosts: the farmers and their six young children.

One student wrote in the diary: "I no longer want to join the army. Yes, I have to defend my country, but isn't that what every soldier thinks? Isn't that what the Germans thought, too?"

Another student wrote: "How can it be that the men got up in the morning, drank their coffee, kissed their wives and children good-bye, and then went to work—work that meant degrading and killing people?"

When the students arrive in Krakow, they are exhausted. They have seen little of the country itself. To them, Poland is first and foremost a collection of killing sites.

Remembrance is one of the key values in Judaism. *Zachor*— "Remember!"—it says in the Torah. How the Holocaust and its victims are remembered, however, has changed in the years since the state of Israel was founded in 1948.

By 1949, 350,000 Shoah survivors had come to Israel. They were received with reservation. In the words of Israeli historian Moshe Zimmermann, they were regarded as "lambs that had allowed themselves to be led to the slaughter." The newly established state of Israel needed heroes and warriors, not victims.

The trauma experienced by the survivors became a taboo: In Israel, their suffering was barely mentioned in public.

The Israeli newspaper *Haaretz* once even declared: "Let's face it: The few Jews we have left in Europe are not necessarily the crème de la crème of our people." Historian Tom Segev discovered that "the Jews in Palestine were fixated on the idea that only the worst elements of society could have survived the camps, in other words, those who would steal the bread of their fellow inmates, etc. The good ones, they thought, had all been killed."

The trial of Adolf Eichmann in Israel in 1961 provided a turning point. Eichmann was responsible for the mass deportation of the European Jews to ghettos and concentration camps. The Israeli chief prosecutor not only put forward files and papers, but also, most notably, summoned many eyewitnesses who spoke openly of their pain and suffering for the first time. According to Tom Segev, the trial "released a whole generation of survivors" and served as "some kind of national group therapy."

Remembering the Shoah became a national purpose, a central, identity-establishing element of the state of Israel, which has lost none of its importance today. Just like elementary and high school teachers, caregivers in Israeli preschools and kindergartens are encouraged to educate children age-appropriately on the Shoah.

The Israeli Ministry of Education also developed a program around school trips to Poland. Since 1988, tens of thousands of children have taken part.

Great care is taken to prepare the teenagers for these journeys, which are not compulsory. When Anat and the other parents discussed whether it is right to expose the children to these horrors, some decided against the trip.

Anat wasn't worried that her son Kai would return from Poland frightened and disturbed. "My worry was rather that he would end up hating the Germans and seeing himself only as a victim of persecution."

On their bus to Krakow, the students rewatched the film *Schindler's List*. The image of Ralph Fiennes as the cruel killer Amon Goeth will still be fresh in their minds when they hear Jennifer Teege's story in Płaszów.

That's why it was so important to Anat that her friend come to Płaszów: "It's too easy to hate Amon Goeth. If the Germans and their allies can turn into murderers, then we, too, can become murderers. If the Germans could turn a blind eye, it can happen to us. I hope that my sons will always remember that. I hope that they will always see the Palestinians as human beings, not as enemies."

Upon their arrival in Krakow, the Israeli teenagers are escorted by three security guards: a bodyguard who has traveled with them from Israel, and two local policemen. They are even more watchful than usual: Four days ago, in Bulgaria, a suicide bomber attacked a passenger bus carrying Israeli tourists; six people died. The Lebanese Hezbollah militia is suspected to have been behind the attack.

The security guards check the bus every time before the children get on board; they inspect every accommodation in advance. The hotel meets all the security requirements: Each level can only be accessed through key-card-operated doors. The rooms for the Israelis are all on the same floor.

Once the teenagers have moved their bags into their rooms, they meet in the hotel lobby, where they rest for a while on modern, pastel sofas between artificial palm trees.

The boys and girls are whispering and giggling; they are planning secret meetings in their rooms. Just like on any ordinary school trip. The teenagers who live on kibbutzim are easily spotted: They walk around the stylish hotel lobby with its dark floor tiles in bare feet. A few men in suits look at them, irritated.

Kai, too, is walking around barefoot. He is the only student in the group who knows that Jennifer Teege will join them in Płaszów. Other than him, only the teachers and the accompanying parents have been told.

■　■　■

NIGHT HAS FALLEN BY THE TIME I enter the hotel in Krakow.

The following day, I meet the Israelis in town. Anat gives me a long hug.

The students have already been to the former ghetto in Podgorze and the Jewish quarter of Kazimierz with its synagogues. Now, for the first time on their trip, they have some free time in the center of Krakow. They want to shop in the Rynek, the historical market square, for souvenirs and gifts for those at home. The trip to the Płaszów memorial is planned for the afternoon.

While the students are walking around by themselves, Anat and I find a secluded café to talk. I tell her about Gerhard, his death. She is grateful that I was still able to come. I tell her that I'll have to return to Munich the next day to prepare for Gerhard's funeral with Inge and my brothers. I want to say good-bye to Gerhard, want to see him in the mortuary chapel before his body is cremated.

Anat shares my pain. She knew Gerhard; she met him when he came to see me in Israel and at my wedding. Anat says that she took to my adoptive father, but that she found the discussions he wanted to have with her, about Hitler and the Third Reich, exhausting.

Even on his deathbed, Gerhard was still thinking about the Holocaust. He wanted to talk to me about Adolf Eichmann; he had read a few books about his trial.

For many years, Gerhard looked into everything related to the Nazis. He did research, read historical sources, compared the numbers of victims. He probed deeper wherever he thought he'd discovered inaccuracies or contradictions.

He became deeply absorbed in the subject. Within the family it played only a minor role, but he had heated discussions with his friends about it—so much so that some of those friendships were destroyed.

Before he died, I suggested to him that he make up with some of those friends. But he didn't want to apologize. My brothers and I couldn't understand why he was so unrelenting, right up to the end.

On his sickbed, he also spoke about Amon Goeth. He quoted Dostoyevsky and asked if man was evil. We discussed the theses put forward by Alexander and Margarete Mitscherlich, who claimed that most Germans were unable to feel guilt or shame after the war. Again, I was unsure what Gerhard was driving at.

I don't think that he could quite put his finger on what exactly was bothering him. He hid behind quotations and theories. But ultimately, it boiled down to his parents. It was all about his own childhood, his mother and his father.

They weren't party members, but were sympathizers and followers. They liked the discipline and the Hitler Youth, and they believed in the secure future that Hitler promised. Opa Bochum regarded the Nazis' success as a blessing for Germany.

In his dying days, Gerhard spoke for the first time at length about his parents: Oma and Opa Bochum.

* * *

Matthias Sieber, the elder of Jennifer Teege's adoptive brothers, believes that, in many discussions, his father was subconsciously trying to defend his parents. Gerhard Sieber was plagued by the question of whether the German people knew about the extermination camps during World War II: "My father knew that the deportations and the disappearance of so many people could not have gone unnoticed. He was asking himself whether his parents had been aware of anything."

Gerhard's parents were keen on Hitler's ideas, and enamored of the man himself: "In the nineteen-fifties, my grandfather once said to my father that Hitler wasn't dead at all and was sure to come back soon. Later, my father regretted not asking his father what he meant by that."

Gerhard Sieber's father died young. Gerhard later tried to talk to his mother about the Nazi years. She claimed not to have known anything about the killing of the Jews.

She once told Matthias that one of the reasons behind the anti-Semitism was the fact that the Jews had owned all the department stores before the war.

Matthias thinks: "My father carried around this unresolved conflict with his parents. He grappled with the Holocaust without realizing that what he really wanted to understand was his parents."

* * *

THE TEACHERS ARE CALLING THE CHILDREN to get ready to leave. We board the bus and drive to the site of the former Płaszów concentration camp.

I am sitting next to one of Kai's classmates. I don't know her; she doesn't say anything. A few others are eyeing me with curiosity. They don't yet know who I am; I am still traveling incognito. Anat and the teachers thought it would best to wait until we got to the memorial before telling them who I am. I lean back and close my eyes; I can rest now before the official part begins.

I don't have any expectations for the day. I only know what I don't want to do, and that's to impart knowledge. That's a job for schoolteachers and university professors. Facts are important; we need them to gain a deeper understanding of things. But if nothing else follows upon the facts, if they are not connected to anything real, nor reflected upon, then they are of little value. They are forgotten as soon as they are heard.

The Israeli students have learned much about the victims on this school trip, and also something about those on the other side, the perpetrators. They must be asking themselves: How could it have happened that some people killed millions of other people?

I want to tell the story from a different perspective. I want to tell them what it is like to be the granddaughter of a concentration camp commandant. And I want to tell them about my relationship with Anat.

Neither Anat nor I knew about my family history when we met by chance. She is a descendant of the generation of victims, I am a descendant of the generation of perpetrators. Nevertheless, our relationship is not symbolic; it is a true friendship that has lasted to this day.

The bus stops by the side of the highway that runs along the boundary of the old camp, and we get off. Once again, I walk up the hill toward the memorial.

The last time I came here, I didn't know what to do with the new information about my family. But I felt there had to be something useful about knowing the horrific truth. I had lived half a life unaware of my origins, but now I had the truth at last. The knowledge shocked me, but it also released me.

Family secrets are corrosive. How often did I despair, feeling like I'd arrived at yet another locked door?

Discovering my family's secret pulled the rug out from under my depression. I felt better after my first trip to Krakow. Today, my sadness is gone.

The first time I came to Krakow, I was still hoping to see my mother again and to build a new relationship with her. It didn't happen. I found her and lost her again.

But I still have my adoptive family.

For a long time, I struggled with my adoptive parents. More than anything, I noticed our differences—everything that separated us. But as Gerhard was dying, I realized how much we have in common. We have spent so many years together and have shared so many experiences. I belong to this family now.

During Gerhard's illness, we were there for each other. It was a wonderful feeling, to be part of a family. When we were planning Gerhard's seventieth birthday at the hospital, Matthias and I went to the house in Waldtrudering and retrieved Oma Bochum's old coffee set and the matching tablecloth from the basement. It has a flower motif, which we all—my adoptive father, his sister and foster siblings, and also my brothers and I—have known since our childhood days.

When we laid the table at the hospital, all the generations present remembered that coffee set. To any outsider, it would have been just some old-fashioned crockery and a patterned tablecloth.

We arrive at the memorial. The students sit down on the steps; the teachers, Anat, and I remain standing before them. One of the teachers says a few introductory words about the Płaszów camp and about its commandant, Amon Goeth.

Next, Anat begins to speak. She describes how I showed up one day at her shared apartment in Tel Aviv that day over twenty years ago, how our friendship has grown and endured to this day. I am touched by her speech.

Anat passes the microphone on to me. I greet the students with "shalom" and go on to describe how I grew up, how I only recently came to discover my family secret. I explain why I stopped writing to Anat and how glad I am to be here with her today. I don't find it hard to tell my story. I tell the students that I will be happy to answer any questions they might have. I want a dialog with them. I don't want to give a lecture; I want to learn something new myself.

■　■　■

At first, some of the students are inattentive as they stand in front of the Płaszów memorial. There is nothing to see but the monument and the green grass. A few of them are already thinking about the next item on the agenda: There will be folk music tonight, something cheerful at last—Polish dancers in traditional costumes, the women wearing flowers in their hair, the men with pointed hats on their heads. The Israeli teenagers will clap and dance, too. The following morning they will be heading to Auschwitz, a sad ending to their trip. Some are afraid of going to Auschwitz.

When Jennifer Goeth starts to explain that she is the granddaughter of Amon Goeth, the students perk up. Some nudge the person next to them: *Who is she? Amon Goeth, how? But she's got dark skin, and she lived in Israel. How can that be?* Many students look shocked, some begin to cry and wipe away their tears with their sleeves. One boy quickly puts on his sunglasses.

The children have many questions. "Was your grandmother a Nazi, too? How did she live in the camp?" "Are you in contact with neo-Nazis?" "How do you cope with it all?" The students' biology teacher wants to know: "Are you afraid of your genes?"

A petite girl with long, dark curls remarks that she knows a lot about the second- and third-generation descendants of the victims, but nothing about their persecutors' descendants: "When Jennifer told us her story, I understood that she and her family are also scarred, in their own way. That she also suffered a trauma." Before she went to Poland, the girl's parents had told her: "Even though everything will be very sad, retain your belief in the good in mankind, in everyone." It's these words that come to her mind now.

Another girl says that she felt an obligation to be touched, to be moved on this trip, which sometimes was a burden: What if she didn't feel anything? But the girl acknowledges: "Jennifer's story, that moved me."

For each commemorative site they are visiting, a group of students has prepared a little ceremony: writing speeches, picking songs, and choosing the color of the flowers. In Płaszów, the ritual is about to begin: One of the young Israelis puts on his yarmulke and gets out his guitar.

Dusk is falling. Joggers are running past; people are walking their dogs. The three security guards accompanying the class have spread out over the area of the former Płaszów camp, climbing the small hills in order to get a clear overview. They coordinate via cell phones while the students prepare for the ceremony, sheets of paper with their speeches at the ready.

■ ■ ■

IT WAS GOOD TO SPEAK in front of the students. They listened spellbound; no one was distracted. I looked into their faces and saw their eyes widen, saw them connecting the past to the present. Afterward, they bombarded me with questions; they wanted to know how I am doing now.

Now they walk up to the memorial; the ceremony begins. Some students read their own compositions in Hebrew; others read testimonies from Płaszów survivors. Then one of the girls sings a

The Israeli students at the Płaszów memorial, in 2012

song, accompanied by one of the boys on guitar. I stand on the edge with Kai and Anat, listening.

Suddenly, one of the Israeli girls by the memorial waves me over: She invites me to take part in their ceremony commemorating the Płaszów victims. I walk to the front; the students take me into their midst. The girl who was going to lay down the flowers gives me a hug. She hands me the bunch of red roses and asks me to lay them down for the class.

I am surprised. At first I hesitate and whisper "no." I am touched by the students' gesture, but I am not sure if it is right for me to perform the ritual. If I am the right person to do it.

The first time I came to Krakow, I brought my own flowers with me and laid them down in private. This time, it is better. This time, I am not alone.

I pause for a second. Then I step forward. I stop in front of the memorial and slowly place the flowers on the stone. And then we sing "Hatikvah," the Israeli national anthem.

Hatikvah means "hope."

BOOKS, FILMS, AND ONLINE

About Amon Goeth, Ruth Irene Goeth, and their daughter Monika

BOOKS:

Awtuszewka-Ettrich, Angelina. "Płaszów – Stammlager," in *Der Ort des Terrors: Geschichte der nationalsozialistischen Konzentrationslager,* Bd. 8 [The place of terror: History of the Nazi concentration camps, Vol. 8]. Edited by Wolfgang Benz and Barbara Distel, 235–87. Munich: Verlag C. H. Beck, 2008.

Crowe, David. M. *Oskar Schindler: The Untold Account of His Life, Wartime Activities, and the True Story Behind the List.* Berkeley: Westview Press, 2004.

Keneally, Thomas. *Schindler's List.* New York: Simon & Schuster, 1994. Originally published as *Schindler's Ark* (London: Hodder & Stoughton, 1982).

Kessler, Matthias. *"Ich muß doch meinen Vater lieben, oder?" Die Lebensgeschichte von Monika Göth, Tochter des KZ-Kommandanten aus "Schindlers Liste"* [I have to love my father, don't I? The life story of Monika Goeth, daughter of the concentration camp commandant from *Schindler's List*]. Frankfurt: Eichborn Verlag, 2002.

Pemper, Mietek. *The Road to Rescue: The Untold Story of Schindler's List.* Translated by David Dollenmayer. New York: Other Press, 2011. Originally published in German as *Der rettende Weg: Schindlers Liste. Die wahre Geschichte.* (Hamburg: Hoffmann und Campe Verlag, 2005).

Sachslehner, Johannes. *Der Tod ist ein Meister aus Wien: Leben und Taten des Amon Leopold Göth* [Death is a master from Vienna: The life and deeds of Amon Leopold Goeth]. Vienna: Styria Premium, 2008.

Segev, Tom. *Soldiers of Evil: The Commandants of the Nazi Concentration Camps*. Translated by Haim Watzman. New York: McGraw-Hill, 1988. Originally published in Hebrew as *Hayale ha-resha* (Jerusalem: Domino Press, 1987).

FILMS:

Blair, Jon. *Schindler: The Real Story*. HBO Home Video, 2001. Originally *Schindler: Die Dokumentation* [Schindler: The documentary]. UK, 1983. PolyGram-Video 1993.

Kessler, Matthias. *Amons Tochter* [Amon's daughter]. Germany, 2003.

Moll, James. *Inheritance*. Allentown Productions, 2006. Film: pbs.org/pov/inheritance/; Panel discussion, 2008: pbs.org/pov/inheritance/video_panel.php.

Spielberg, Steven. *Schindler's List*. Universal, 1993.

Ze'evi, Chanoch. *Hitler's Children*. Israel, 2011. hitlerschildren.com.

WEB:

A series of interviews by Mietek Pemper with Monika Hertwig: mietek-pemper.de/wiki/Interview_mit_Monika_Hertwig.

Selected reports of Płaszów survivors

Frister, Roman. *The Cap: The Price of a Life*. Translated by Hillel Halkin. New York: Grove Press, 1999. Originally published in Hebrew as *Deyokan 'atsmi 'im tsaleket* (Tel Aviv: Devir, 1993).

Müller-Madej, Stella. *A Girl from Schindler's List*. Translated by William R. Brand. London: Polish Cultural Foundation, 1997. Originally published in German as *Das Mädchen von der Schindler-Liste* (Augsburg: Olive Tree Publishing, 1994).

BOOKS:

Brunner, Claudia, and Uwe von Seltmann. *Schweigen die Täter, reden die Enkel* [When the perpetrators keep silent, their grandchildren will talk]. Frankfurt: Fischer Verlag, 2006.

Frank, Niklas. *In the Shadow of the Reich*. Translated by Arthur S. Wensinger. New York: Knopf, 1991. Originally published in German as *Der Vater: Eine Abrechnung* (Munich: Verlag C. Bertelsmann, 1987).

Frank, Niklas. *Meine deutsche Mutter* [My German mother]. Munich: Verlag C. Bertelsmann, 2005.

Himmler, Katrin. *The Himmler Brothers: A German Family History*. Translated by Michael Mitchell. London: Macmillan, 2007. Originally published in German as *Die Brüder Himmler: Eine deutsche Familiengeschichte* (Frankfurt: S. Fischer Verlag, 2005).

Lebert, Norbert, and Stephan Lebert. *My Father's Keeper*. Translated by Julian Evans. Boston: Little, Brown, 2001. Originally published in German as *Denn Du trägst meinen Namen: Das schwere Erbe der prominenten Nazi-Kinder* (Munich: Karl Blessing Verlag, 2000).

Nissen, Margret. *Sind Sie die Tochter Speer?* [Are you Speer's daughter?]. Munich: Deutsche Verlags-Anstalt, 2004.

Saur, Karl-Otto, and Michael Saur. *Er stand in Hitlers Testament: Ein deutsches Familienerbe* [He was named in Hitler's will: A German family legacy]. Berlin: Econ Verlag, 2007.

Senfft, Alexandra. *Schweigen tut weh: Eine deutsche Familiengeschichte* [Silence hurts: A German family history]. Berlin: Claassen Verlag, 2007.

Schirach, Richard von. *Der Schatten meines Vaters* [My father's shadow]. Munich: Carl Hanser Verlag, 2005.

Timm, Uwe. *In My Brother's Shadow*. Translated by Anthea Bell. London: Bloomsbury, 2005. Originally published in German as *Am Beispiel meines Bruders* (Cologne: Verlag Kiepenheuer & Witsch, 2003).

FILM:

Ludin, Malte. *2 or 3 Things I Know About Him*. National Center for Jewish Film, 2007. Originally *2 oder 3 Dinge, die ich von ihm weiß* (Germany, 2005), absolut Medien, 2oder3dinge.de.

About the psychology of perpetrators and followers, of second and third generations of descendants of Nazi criminals, plus general information about trauma therapy

Bar-On, Dan. *Legacy of Silence: Encounters with Children of the Third Reich*. Cambridge: Harvard University Press, 1989.

Kellner, Friedrich. *Vernebelt, verdunkelt sind alle Hirne: Tagebücher 1939– 1945* [They see no evil, hear no evil: Diaries 1939–1945]. Edited by Sascha Feuchert, Robert Kellner, Erwin Leibfried, Jörg Riecke, and Markus Roth. Göttingen: Wallstein Verlag, 2011.

Kogan, Ilany. *Der stumme Schrei der Kinder: Die zweite Generation der Holocaust-Opfer* [The silent scream of the children: The second generation of Holocaust victims]. Frankfurt: S. Fischer Verlag, 1998.

Mitscherlich, Alexander and Margarete Mitscherlich. *The Inability to Mourn: Principles of Collective Behavior*. Translated by Beverley R. Placzek. New York: Grove Press, 1975. Originally published in German as *Die Unfähigkeit zu trauern: Grundlagen kollektiven Verhaltens* (Munich: Piper Verlag, 1967).

Ruppert, Franz. *Trauma, Bonding & Family Constellations: Understanding and Healing Injuries of the Soul*. West Sussex: Green Balloon Publishing, 2008. Originally published in German as *Trauma, Bindung und Familienstellen: Seelische Verletzungen verstehen und heilen*. (Stuttgart: Klett-Cotta Verlag, 2007).

Welzer, Harald, Sabine Moller and Karoline Tschuggnall. *"Opa war kein Nazi": Nationalsozialismus und Holocaust im Familiengedächtnis* [Grandpa wasn't a Nazi: National Socialism and the Holocaust in family memory]. Frankfurt: S. Fischer Verlag, 2002.

Welzer, Harald. *Täter. Wie aus ganz normalen Menschen Massenmörder werden* [Perpetrators. How ordinary people become mass murderers]. Frankfurt: S. Fischer Verlag, 2005.

Westerhagen, Dörte von. *Die Kinder der Täter: Das Dritte Reich und die Generation danach* [The children of the perpetrators: The Third Reich and the next generation]. Munich: Kösel Verlag, 1987.

Yalom, Irvin D. *Staring at the Sun: Overcoming the Terror of Death*. San Francisco: John Wiley & Sons, 2008. Published in Germany as *Der Panama-Hut oder Was einen guten Therapeuten ausmacht* (Munich: Goldmann Verlag, 2002).

Other

Nattiv, Guy. *Mabul (The Flood)*. Scripted by Noa Berman-Herzberg and Guy Nattiv. Israel: K5 International Producers, 2011.

Special Note

If you are interested in looking into your own family history of the Nazi era, you can find exceptional guidance here (in German): https://chrismon.evangelisch.de/comment/21672.

IN *MY GRANDFATHER WOULD HAVE SHOT ME*, Jennifer Teege searches for answers about her biological family's haunted past. She also gives readers an opportunity to revisit one of the darkest periods of the twentieth century and consider its still-reverberating consequences. This guide takes a closer look at the key people, conflicts, and themes at the heart of Jennifer Teege's story—such as family, race, friendship, adoption, and the keeping of secrets—as well as the Holocaust and its aftermath.

1. The *Washington Post* describes this book as equal parts "a memoir, an adoption story, and a geopolitical history lesson." As Jennifer Teege researches and reflects on the Goeth family, the larger history of Nazism, and her own adoption and childhood, which of these do you think affects her the most? Which was most interesting to you, and why?

2. In the opening of the book, Jennifer mentions that she once identified as Jennifer Goeth (1). How does this observation about her name lay the groundwork for Jennifer's exploration of her identity?

3. The book is presented from intertwined perspectives: first-person from Jennifer and third-person from Nikola Sellmair. Discuss how the two authors complement and complicate one another's views.

4. Jennifer has trouble reconciling her memory of a grandmother she loved with the truth of a woman who lived with Amon Goeth and ignored his atrocities. What does Jennifer's attitude toward Ruth, and Ruth's toward Goeth, suggest about love? Is it possible to love one part of a person while rejecting another part?

5. Nikola observes: "[Demonizing the prominent Nazis] offers a way out of having to deal with one's own actions, one's family's actions—or indeed those of the many people who joined in on a small scale" (39 – 40). Are those who join on a "small scale" responsible for the actions of the whole? Why or why not? Can you connect your answer to one or more contemporary social issues?

6. Monika Goeth, named for a father she never met, belongs to the first generation of descendants of Nazi perpetrators. For her, it was "Goeth's story that shaped her identity" (99). How was her experience with the family history different from Jennifer's? How will that experience change for Jennifer's children? Discuss the impact of family trauma as it is passed down through generations.

7. In an interview, Jennifer has said, "Today I see [Monika] not only as my birth mother, but also as a woman with her own story and history. She suffers from the weight of the past." What do you think motivated Monika to put Jennifer up for adoption, and to conceal the family secret from her?

8. A secret is often kept by more than one person. Peter Bruendl, Jennifer's psychoanalyst, identifies this phenomenon as a "conspiracy of silence" (17). Discuss the possible repercussions of keeping a secret, and whether doing so can be justified.

9. After living in Israel, Jennifer has strong ties to its culture and people. She feels guilty about her family history and is reluctant to share it with her Israeli friends. Discuss her fears and her friends' reactions. How would you have acted in Jennifer's place, and reacted as her friend?

10. Jennifer's African heritage comes up many times in the narrative. Jennifer describes her skin color as a "barrier" (41) between her and Goeth and as "good camouflage" (175) in Israel. Describing her first visit to the African quarter of Paris, she says, "It was a strange world to me, but at the same time I had a sense of homecoming" (184). What role does race play in Jennifer's quest for identity?

11. While researching concentration camp commandants, historian Tom Segev interviewed close relatives of many Nazi perpetrators. Segev found that most relatives alter their memories or forget events altogether (83). Discuss the relationship between history and memory.

12. On crossing the boundary between fiction and history, Jennifer writes, "Slowly I begin to grasp that the Amon Goeth in the film *Schindler's List* is not a fictional character, but a person who actually existed in flesh and blood" (7). Think about how your view of the Holocaust has been shaped by popular culture. How has this book informed your understanding of the Holocaust and its aftermath?

Q: Your true story reads like a perfect storm of unlikely circumstances. Your mother is German, and your father is Nigerian. You were adopted at age 7 by a white family, but still had intermittent contact with your birth mother and grandmother. You went to college in Israel, where you have many Jewish friends to this day, and you speak fluent Hebrew.

Then, at age 38, you picked up a library book by chance and made this stunning discovery: Your grandfather is Amon Goeth, the infamous Nazi commandant of Płaszów concentration camp. Even more, the woman you remember as your kindly grandmother lived with him in luxury at that very camp. And your birth mother, Monika—who kept these secrets from you—appeared in a documentary on the subject, which premiered the day after you found that book. Looking back, is there any one thing that was hardest for you to accept?

JENNIFER TEEGE: There was no one point that was hardest to reconcile with. I have faced many challenges in life that I've had to overcome. It started with my birth: For any child, being given away by your parent is a traumatic experience that you carry with you for a lifetime. One's self-worth is called into question. My mother was a single parent, and she placed me in a Catholic home for children when I was four weeks old.

In my case, another traumatic experience was added to the first. When I found out who my grandfather was from the book about my mother, my initial reaction was shock—first, shock that there was a book about my mother that I hadn't been aware of, and next, shock that Amon Goeth was my grandfather. Like so many others, I had seen the movie *Schindler's List*. I was terrified of that monstrous figure—and now I was petrified: Was it possible that I might resemble him? Today that fear is gone. I know that I am a different person.

Q: Following your discovery, you fell into a deep depression. Many descendants of Holocaust victims are haunted by their family's past. Is it different for descendants of perpetrators?

JT: Victims' and perpetrators' families are both bound by silence, by the desire to forget. In Germany the Holocaust is taught in detail at school but seldom discussed within the family. Many people of my generation are afraid to ask exactly what their own grandparents did during the war, how they were involved. They are afraid of the truth. In victims' families, the dynamic is somewhat different. Most don't want to ask because they don't want to traumatize the survivors further, or again. One might say that they respect the silence of the survivors.

Q: Did your discovery impact how you see yourself? Did it make you feel differently about your own children?

JT: Once I compared my life to a puzzle. There were so many pieces, but the frame was missing. I was unsure of where I fit in, and where my depression stemmed from. Today I can put the pieces into a frame, and they make up a clear picture—things aren't in a big mess anymore. Of course there are still blind spots; everyone has them. The idea that I'm going to find *all* the pieces by the end of my life is unrealistic, but it's also not important. Now I have a different way of assessing my feelings. Even if I have changed, I've remained the same person—I know who I am. I can live with the past now.

My children can, too. They are a whole other generation removed; for them it's not their grandfather, it's their great-grandfather. They are shielded, in part, by that greater distance.

Q: Even while grappling with depression, you extensively researched your grandfather and visited the sites of his crimes. Later on, you reconnected with your biological mother. What were you hoping to achieve?

JT: The questions started piling up in my head. I had to work through them bit by bit. First of all I had to confront my anger and disappointment toward my biological mother: Why did she reject me? It took a long time for my image of her to change. Today I see her not only as my birth mother, but also as a woman with her own story and history. She suffers from the weight of the past. She once said, "I live with the dead."

I wanted to let go, to live in the here and now again, and be a good mother to my own children. On my path in that direction, Krakow was an important step. I realized that I am part of the third generation. I know now that I am not to blame, and the guilt no longer weighs heavily on my shoulders. That said, today I am occupied with the concept of responsibility. Everyone bears a responsibility to add value to their surroundings. I carry responsibility not only as a German woman, or as Amon Goeth's granddaughter, but simply as a person.

Q: You waited a long time to tell your friends in Israel about your discovery. Why did you break off contact? When were you ready to approach them again?

JT: It was over two years before I could bring myself to reveal my Nazi connection to my Jewish friends. First, I wanted to come to grips with it myself, so as not to confront them with too many details.

Earlier, I addressed the sorrow in the families of Holocaust victims, the inability to cope with old fears and feelings. I didn't know exactly what my friends' stories were: Where were their relatives born? When were they killed, and how did they die? Was there a connection to Poland, to Płaszów? So I was afraid of what my revelation would trigger for them. But my friends reacted in the most empathetic way imaginable—they cried with me.

Q: Amon Goeth—your grandfather—is the brutal Nazi commandant played by Ralph Fiennes in *Schindler's List*. When did you first see the movie? How did it affect you?

JT: Many assume that I have watched *Schindler's List* over and over. But I haven't; I'm not sure why. I just don't feel the need to. I did see it when it was first released—in the mid-'90s, while I was living in Israel. At the time, it was just a film to me; it didn't have anything to do with me personally. I wasn't suspicious when I heard the name Goeth, in part because the movie uses the Polish pronunciation, which is different than the German, with a shorter "th." The movie touched me, but I was very conscious that it was a Hollywood feature, not a documentary.

Q: Many would say that your story is almost unbelievable, "stranger than fiction." In your memoir you say, "At age 38, I found *the book*. Why on earth did I pick it up off the shelf, one among hundreds of thousands of books? Is there such a thing as fate?" You also say, "I've realized that the book is meant for me, the key to my family history, to my life. The key I've been looking for all these years." What do you make of the coincidences in your past?

JT: It depends on how you view life: forward-thinking or backward-thinking. When I first went to Israel in my mid-twenties, I didn't go because I was German and wanted to make amends. No, I originally went for a vacation, fell in love, and stayed. I never dreamed that there was a close tie between the country and myself. Today, I know that the stay was fateful for me.

I like to use the tree as a metaphor. Everyone has their own genealogy, with roots, a trunk, and branches. We can develop and thrive in this system, but we can't direct it. We all exist within a larger pattern that carries and sustains us.

Q: Your story is a bestseller in several languages. After seventy years, why do you think there remains such a strong interest in the Holocaust? And how does your story speak to readers, beyond history?

JT: The Holocaust was a genocide of gigantic proportions. That makes it so difficult to grasp. Why did so many people support the Nazi ideology, stay silent, look away, or participate? Why do people kick other people down? How do people turn into perpetrators? Our world is still struggling with the same themes: racism, extremism, and violence. My book doesn't claim to have the answers; it seeks only to encourage contemplation. Along with the historical connections I'm interested in psychology—the toxic power of family secrets, depression and how to deal with it, and of course the search for one's own identity. The book speaks to readers on all of these points in a single story.

Q: Why was it important that you share your story? What do you hope readers will take away?

JT: First of all, I didn't write my story down with some specific, concrete goal in mind. But I thought that it was important to share my story in part because of a quote I read years ago, from Bettina Goering, another descendant of a perpetrator. She said that she and her brother had had themselves sterilized, so as not to produce any more Goerings. I think that that attitude sends the wrong message. There is no Nazi gene: We can decide for ourselves who and what we want to be.

Questions provided by the publisher.

ABOUT THE AUTHORS

JENNIFER TEEGE has worked in advertising since 1999. Teege studied Middle Eastern and African studies in Israel, where she also learned fluent Hebrew. Teege lives in Germany with her husband and two sons. This is her first book.

For more information about Jennifer Teege, please visit jennifer-teege.com. She is available for speaking engagements. Please contact speakersbureau@workman.com.

NIKOLA SELLMAIR graduated from Ludwig-Maximilians-University Munich and has worked in Hong Kong, Washington, D.C., Israel, and Palestine. She has been a reporter in Munich at Germany's *Stern* magazine since 2000. Her work has received many awards, including the German-Polish Journalist Award, for the first-ever article about Jennifer Teege's singular story.

■ ■ ■

ABOUT THE TRANSLATOR

Born in Germany, CAROLIN SOMMER studied applied languages at universities in Germany, France, and the UK before settling in the UK in 1997. After working in bilingual roles for multinational companies for several years, she took a career break to look after her three sons. Translations have been her career focus since 2011.